# Indonesia
## the beyond
# Water's Edge

The Research School of Pacific and Asian Studies (RSPAS), a part of the ANU College of Asia and the Pacific at The Australian National University, is home to The Indonesia Project, a major international centre, which supports research activities on the Indonesian economy and society. Established in 1965 in the School's Division of Economics, the Project is well known and respected in Indonesia and in other places where Indonesia attracts serious scholarly and official interest. Funded by the ANU and the Australian Agency for International Development (AusAID), the Indonesia Project monitors and analyses recent economic developments in Indonesia; informs Australian governments, business and the wider community about those developments and about future prospects; stimulates research on the Indonesian economy; and publishes the respected *Bulletin of Indonesian Economic Studies*.

The School's **Department of Political and Social Change** (PSC) focuses on domestic politics, social processes and state–society relationships in Asia and the Pacific, and has a long-established interest in Indonesia.

Together with PSC and RSPAS, the Project holds the annual Indonesia Update conference, which offers an overview of recent economic and political developments and devotes attention to a significant theme in Indonesia's development. The Project's *Bulletin of Indonesian Economic Studies* publishes the economic and political overviews, while the proceedings related to the theme of the conference are published in the Indonesia Update Series.

The **Institute of Southeast Asian Studies (ISEAS)** was established as an autonomous organization in 1968. It is a regional centre dedicated to the study of socio-political, security and economic trends and developments in Southeast Asia and its wider geostrategic and economic environment. The Institute's research programmes are the Regional Economic Studies (RES, including ASEAN and APEC), Regional Strategic and Political Studies (RSPS), and Regional Social and Cultural Studies (RSCS).

**ISEAS Publishing**, an established academic press, has issued almost 2,000 books and journals. It is the largest scholarly publisher of research about Southeast Asia from within the region. ISEAS Publishing works with many other academic and trade publishers and distributors to disseminate important research and analyses from and about Southeast Asia to the rest of the world.

Indonesia Update Series

# Indonesia
## beyond
## the beyond
# Water's Edge
## Managing an
## Archipelagic State

EDITED BY ROBERT CRIBB
AND MICHELE FORD

**ISEAS**

**Institute of Southeast Asian Studies**
**Singapore**

First published in Singapore in 2009 by
ISEAS Publishing
Institute of Southeast Asian Studies
30 Heng Mui Keng Terrace
Pasir Panjang
Singapore 119614

E-mail: publish@iseas.edu.sg
http://bookshop.iseas.edu.sg

*The responsibility for facts and opinions in this publication rests exclusively with the authors and their interpretations do not necessarily reflect the views or the policy of the Institute or its supporters.*

### ISEAS Library Cataloguing-in-Publication Data

Indonesia beyond the water's edge : managing an archipelagic state / edited by Robert Cribb and Michele Ford.
1.  Archipelagoes — Indonesia
2.  Territorial waters — Government policy — Indonesia
3.  Indonesia — Politics and government — 1988-
I.  Cribb, R. B.
II.  Ford, Michele.
KZ3881 I5I41                    2009

ISBN 978-981-230-984-6 (soft cover)
ISBN 978-981-230-985-3 (hard cover)
ISBN 978-981-230-981-5 (PDF)

Edited and typeset by Beth Thomson, Japan Online, Canberra
Indexed by Angela Grant, Sydney
Printed in Singapore by Utopia Press Pte Ltd

# CONTENTS

# TABLES

# MAPS AND FIGURES

## MAPS

## FIGURES

# CONTRIBUTORS

I Made Andi Arsana is a Lecturer in the Department of Geodesy and Geomatics, Gadjah Mada University, Yogyakarta. He is currently undertaking a PhD at the Australian National Centre for Ocean Resources and Security, University of Wollongong, Wollongong.

Sam Bateman retired from the Royal Australian Navy as a Commodore. He is now a Professorial Research Fellow at the Australian National Centre for Ocean Resources and Security, University of Wollongong, and concurrently a Senior Fellow and Adviser to the Maritime Security Programme at Nanyang Technological University, Singapore.

John G. Butcher is an Associate Professor in the Department of International Business and Asian Studies, Griffith University, Brisbane. He is currently conducting research with Robert Elson on the history of claims to maritime territory in the Indonesian archipelago since 1850.

Robert Cribb is Professor of Indonesian History at the Australian National University, Canberra. His research focuses on issues of national identity, mass violence, historical geography and environmental politics in Indonesia. His publications include the *Historical Atlas of Indonesia* (2000).

Hasjim Djalal is a Member of the Indonesian Maritime Council; Advisor to the Indonesian Minister for Marine Affairs and Fisheries; and Advisor to the Indonesian Naval Chief of Staff. He helped shape Indonesia's position at the Third United Nations Conference on the Law of the Sea and has served as Indonesia's Ambassador at-large for the Law of the Sea and Maritime Affairs (1994–2000). He has written extensively on domestic, regional and international maritime issues.

**Rili Djohani** joined The Nature Conservancy in 1995 to help establish its coastal and marine program in Indonesia. She was The Nature Conservancy's Country Director for Indonesia from 2004 to 2008 and is now the Director of its Coral Triangle Program. She is also a PhD Scholar at the Van Vollenhoven Institute in the Faculty of Law, University of Leiden.

**Michele Ford** chairs the Department of Indonesian Studies at the University of Sydney, Sydney. Her research interests include the Indonesian labour movement, labour migration and the Riau Islands.

**James J. Fox** is a Professor in the Resource Management in Asia Pacific Program in the Research School of Pacific and Asian Studies, Australian National University. He has been closely involved in the study of the fishermen of eastern Indonesia for nearly two decades.

**Lenore Lyons** is an Associate Professor at the Centre for Asia-Pacific Social Transformation Studies, University of Wollongong. She recently completed an Australian Research Council Discovery project with Michele Ford on transnational encounters between Singaporeans and people living in the Riau Islands.

**Arif Havas Oegroseno** is Director of Treaties for Political, Security and Territorial Affairs in the Indonesian Ministry of Foreign Affairs. He has served as a senior Indonesian diplomat in Portugal and at the United Nations, and has been extensively involved in planning Indonesia's broad maritime security strategy.

**David Ray** is Director of the Indonesia Infrastructure Initiative, a project funded by the Australian Agency for International Development (AusAID) to promote infrastructure development in Indonesia. He was previously Deputy Director of the SENADA project, a program to strengthen the competitiveness of Indonesia's labour-intensive light manufacturing industries.

**Erwin Rosmali** has served as Section Head of Belawan and Tanjung Priok ports, Harbour Master of Makassar port, Division Head of Ship Seaworthiness at Tanjung Priok port and Head of the Subdirectorate of Ship Equipment, Machinery and Radio in the Directorate of Marine Safety, Ministry of Transport.

**Clive Schofield** is a QEII Research Fellow at the Australian National Centre for Ocean Resources and Security, University of Wollongong. His research focuses on the intersection of geographical/technical, legal and political disciplines in the law of the sea, with particular reference to maritime boundary delimitation.

**Djoko Sumaryono** is the Chief Executive of the Indonesian Maritime Security Coordinating Board. He has held a number of key posts in the Indonesian Navy and served as Permanent Secretary to the Coordinating Minister for Political and Security Affairs in 2004.

**Sarah Waddell** is a lawyer who has worked in Indonesia as a consultant in environmental law, coastal resources management and water resources management (2001–08). She has directed a number of legal training programs on legislative drafting, anti-corruption prosecution and environmental law enforcement under the Indonesia Australia Specialised Training Program (IASTP). She is currently Senior Researcher at the Australasian Legal Information Institute (AustLII) working on international projects related to access to the law.

# ACKNOWLEDGMENTS

---

This book is a product of the 26th Indonesia Update Conference, held at the Australian National University (ANU) on 19–20 September 2008. This conference is held annually under the auspices of the Indonesia Project and the Department of Political and Social Change, both in the Research School of Pacific and Asian Studies (RSPAS) at the ANU.

The conference was made possible by the efforts of a strong and invariably cheerful team of organizers—Liz Drysdale, Cathy Haberle, Allison Ley, Anne Looker and Trish van der Hoek—as well as a most helpful advisory committee consisting of Chris Manning, Budy Resosudarmo and Ross McLeod. ANU student volunteers also helped in many ways.

As with the previous annual Indonesia Update conferences, the Australian Agency for International Development (AusAID) provided generous financial and practical support. The conference drew substantially, but intangibly, on the resources and goodwill of the Arndt-Corden Division of Economics and the Division of Pacific and Asian History (ANU).

We are indebted to Beth Thomson for her meticulous copy editing and for other valuable advice on preparing the manuscript. The Cartography Department at RSPAS drew two of the maps.

Robert Cribb and Michele Ford
*Canberra and Sydney*
*July 2009*

# 1 INDONESIA AS AN ARCHIPELAGO: MANAGING ISLANDS, MANAGING THE SEAS

*Robert Cribb and Michele Ford*

Indonesia is the world's largest archipelagic state. By the latest official count, the archipelago consists of 18,108 islands, which lie scattered between the mountainous island of Breueh in the west and tiny Sibir Island in Humboldt Bay (Teluk Yos Sudarso) in the east, and between Miangas in the north and Dana in the south. Indonesia's islands range in size from New Guinea, Borneo and Sumatra, respectively the second, third and sixth largest islands in the world, to tiny islets with only local names (see Map 1.1).[1] Situated between longitude 97°E and 141°E and between latitude 6°N and 11°S, Indonesia comprises 2.8 million square kilometres of water (including 92,877 square kilometres of inland waters) and 1,826,440 square kilometres of land. If Indonesia's exclusive economic zone (EEZ), stretching beyond the archipelago, is included, Indonesia's area of sea expands to 7.9 million square kilometres.

Indonesia's archipelagic character creates two distinct but intertwined problems of governance. First, by separating Indonesia's landmass into islands, the sea creates special challenges of communication, coordination and even identity. Governing the land is made more difficult by the

---

1  Although Indonesia is the world's largest archipelagic state, it is exceeded in the number of islands by Canada's Arctic archipelago, which has 36,463 islands and covers 1.4 million square kilometres; see 'Arctic archipelago', *Canadian Encyclopedia*, http://www.thecanadianencyclopedia.com/, accessed 25 November 2008. Indonesia's coastline, at 54,716 kilometres, is also a distant second to Canada's at 202,080 kilometres. Some 5,707 Indonesian islands possess official names (Kwiatkowska 1991: 14).

Map 1.1    The Indonesian archipelago

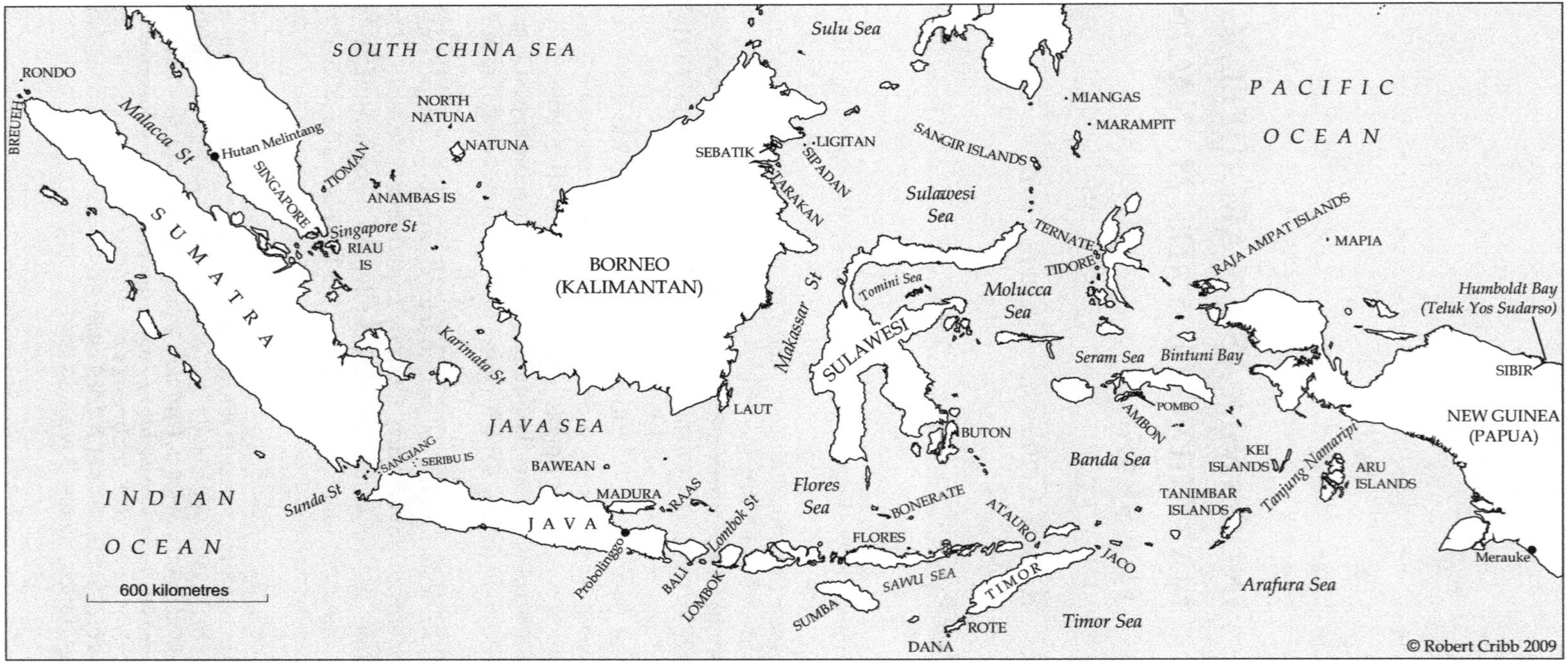

intervening presence of the sea. Second, the seas that lie between and around these islands need to be governed. These seas represent a major strategic, economic and cultural resource for Indonesia; they cannot be ignored, yet governing the maritime zone poses enormous practical difficulties.

## ON BEING AN ARCHIPELAGO

Indonesia's status as an archipelago has important consequences both for its identity as a nation and for its character as a state. Although Indonesia's first president, Sukarno, confidently asserted that even a child could see that the arc of islands between Asia and Australia constituted a single national unit (Sukarno 1961: 11), most observers look at Indonesia's geography and see the potential for fragmentation rather than a self-evident whole. To many people, islands seem destined by their very nature for separate existences and Indonesia's unity consequently has appeared to be fragile, artificial, perhaps even imaginary.[2] The closest geographical analogy to Indonesia, moreover, is a model of fragmentation. The Caribbean archipelago, 7,000-odd tropical islands stretched between two continents, is marked by cultural diversity and a long colonial history. Known to much of the world as the West Indies, it has a culturally diverse population of around 40 million. But it is divided into 27 independent states and dependent territories.

The practical reality is that Sukarno and the sceptics are both right: the sea divides and unites archipelagos in complex ways. The most powerful effect of the seas between the islands of an archipelago, however, is to separate those islands from each other, politically, economically, socially and culturally. The sea divides, because the process of moving people and cargoes safely from the shore onto the high seas is almost always more dangerous, more demanding of technology and more costly of time and effort than any such movement on land. A stretch of intervening sea magnifies the practical distance between two pieces of land. Yet the divisive force of the sea should not be overstated in comparison with that of land. In any large country sheer distance, whether overland or by sea, is divisive. So are differences in topography, climate and even time zone. The problems of governing the scattered islands of Indonesia are not demonstrably greater than those of governing the landmass of the Russian Federation or China.

The sea also unites because, paradoxically, it collapses distance in a way that land does not. That is to say, once a traveller is safely aboard

2 This issue is discussed in Cribb (1999).

and away from the hazards of the coast, the difference in danger, difficulty and cost between a long journey and a short one is much smaller than the comparable difference on land. There is no clearer proof of the advantages of a long, unimpeded journey over one involving multiple disembarkations than the fact that the transit around the Cape of Good Hope became the preferred route for transport between Europe and Asia soon after its discovery, despite the availability of the much shorter passage involving land transport across the Isthmus of Suez. And even today, there is a thriving ferry service between Jakarta and Surabaya, both on the island of Java, despite the availability of an overland connection between the two ports. In Chapter 6 of this volume, David Ray illustrates the critical importance of loading costs to the overall cost of sea transport. The combined effect of the difficulty of going to sea and the collapsing of distance once at sea is that archipelagos, where mastery of sea-going technology is a necessity, tend to be well linked internally despite different distances between islands. These links are the basis for the separate political, economic, social and cultural identities of archipelagos. Sukarno's characterization of the archipelago may have been instinctive and naive rather than analytical, but his insight that archipelagos have a natural unity was entirely apt.

Indonesia's archipelagic character is qualified in two important ways. First, what we conventionally call the Indonesian archipelago is part of a much larger archipelagic assembly stretching between the continents of Asia and Australia and divided between Indonesia, Malaysia, Singapore, Brunei, the Philippines, Timor-Leste and Papua New Guinea. Indonesia even shares four islands—Borneo, Sebatik, Timor and New Guinea—with neighbouring countries. As parts of this larger archipelagic assembly, Indonesia's islands are easily drawn into relationships with the islands of its neighbours. In Chapter 13 of this book, Michele Ford and Lenore Lyons discuss the intimate relationship between the Riau Islands and Singapore. Comparable relations could be identified between Sumatra and West Malaysia, between Indonesian and Malaysian Borneo, and even between northern Sulawesi and the Philippines. The fact that, once one can get off an island and aboard a vessel, the sea offers the possibility of travel in almost any direction has a distinctly centrifugal effect. From a small island in the heart of Indonesia, one can sail directly to Singapore, or Hong Kong, or Port Moresby, or Darwin. The island of Bawean in the heart of the Java Sea, for instance, was a long-term provider of labour to British Singapore in the colonial era, despite its proximity to Java (Encyclopaedie van Nederlandsch-Indië 1917: 212). In contrast, from a small town in the heart of Siberia, one's movement to the outside world is largely constrained by the trajectory of a few roads. Apparently peripheral islands, therefore, have a capacity to be engaged with the wider world that is mostly unavailable to peripheral land-bound regions.

Second, unlike many of the world's archipelagic states, Indonesia is actually a complex of archipelagos and large islands with a single dominant island, Java. In this respect it resembles Britain, Fiji, Japan and New Zealand — and is correspondingly unlike the Philippines, the Maldives and Tuvalu, where the pre-eminence of the main island is much less pronounced. The dominance of Java is not clear from simple geography, but it stands out if we represent the size of Indonesia's regions according to population (see Figure 1.1).

*Figure 1.1    Indonesia's archipelago by population*

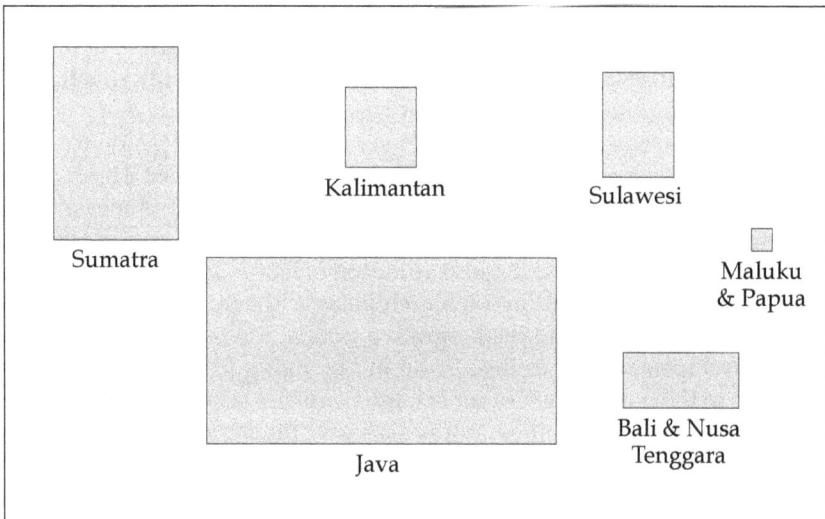

The disparity between the islands in terms of population means that Indonesian politics has historically been marked by a centre–periphery tension that has given rise to one of the most enduring perceived divisions within Indonesia, that between Java and the outer islands (which the Dutch called the *buitengewesten*). Dutch dominance on Java was well established for at least a century and a half before major parts of the other islands came under effective Dutch control. This historical fact, together with Java's preponderance in population, its cultural self-confidence and its formidable economic assets, meant that Java set the model for the colonial governance of the archipelago (Dick 1996: 32). The inclusion in the Netherlands Indies of Sumatra — also with a substantial population and impressive economic resources — meant that the nation that emerged within the shell of colonial rule in the first half of the twentieth century was more than merely Greater Java.

Javanese have always dominated Indonesian politics, but they have never been the defining ethnic group of the Indonesian nation in the way that Han Chinese, Kinh (Viet) and Thai peoples, respectively, define China, Vietnam and Thailand, relegating other ethnic groups to minority status. The emergence of Malay, rather than Javanese, as the lingua franca of the colony and later as the national language of Indonesia is only the most obvious reflection of Java's failure to capture the identity of the new nation. The qualified nature of Javanese dominance enhances the sense that Indonesia's unity is fragile. A dozen or more of the non-Javanese ethnic groups in the archipelago are large enough to claim credibly a separate national status. In contrast, independence for any of the minorities in Vietnam, for instance, is highly implausible. Indonesia's basic ethnic configuration makes separatism a believable political option, but it is the country's archipelagic character that makes it seem actually practicable. For this reason, despite the fact that there have been only four serious separatist movements in Indonesia's six decades of independence — the Republic of the South Moluccas in the 1950s, and the East Timor, West Papua and Aceh separatist movements more recently — Indonesia's rulers have displayed an obsession bordering on paranoia concerning the prospect of separatist movements (Emmerson 2005: 38).

Yet, even though Indonesia's archipelagic character seems, at least potentially, to exacerbate disintegrative trends, the fact of being part of a vast archipelago also unites. Pride in belonging to a great state is an element in the nationalism of most large countries — in China, the United States and Russia as well as Indonesia — and the very fact of inhabiting the world's largest archipelagic state is a source of satisfaction to many Indonesians. This satisfaction is evident in the inflation of the number of islands officially said to comprise the archipelago. In 1963, the number of islands was formally estimated at 13,667; this tally took on an iconic quality and was not influenced by the appearance and disappearance of islands as a result of siltation, erosion, land reclamation and tectonic movements,[3] and not even by the 1976 annexation of the islands of Atauro and Jaco, off the former Portuguese colony of East Timor. In 1994, the figure was revised upward to 17,508. (This total, too, remained unrevised when Indonesia lost Atauro and Jaco to independent Timor-Leste in 1999.) In 2003, shortly after the loss of Sipadan and Ligitan to

---

3   An earthquake in 2007 produced six new islands near South Pagai off the west coast of Sumatra; two islands were lost in the construction of Jakarta's Soekarno-Hatta International Airport. In 2007, the State Ministry for the Environment estimated that Indonesia had lost 24 islands during 2005–07 (Muhammad 2008).

Malaysia under an International Court of Justice ruling, the figure was revised again to 18,108, an island being defined for this purpose as any piece of land surrounded by water and having an area greater than 30 square metres.[4]

Perhaps more important, there is some sense in Indonesia that being an archipelago is precisely what makes the country work, because it puts distance between groups that might otherwise clash. When former president Megawati Sukarnoputri announced her support for a bridge between Bali and Java in 2004, many Balinese opposed it because it would diminish Bali's isolation (Leinbach and Ulack 1999: 221; Pringle 2004: 220–21). Greg Acciaioli has argued that Soeharto's New Order used the myth of a maritime way of life among all the peoples of the archipelago across vast stretches of time to create a sense of identity that was effective because it sidestepped the practical political problems of life on land (Acciaoli 2001: 5–7).

A number of scholars have used the metaphor of archipelago to highlight the way in which dispersed institutions are able to share an identity because the physical distance that separates them is trumped by the social distance that detaches them from their immediate environment but unites them with each other. Alexander Solzhenitsyn (1974) coined the term 'Gulag archipelago' for the scattering of political prison camps across the face of the Soviet Union. These islands of incarceration shared a way of life and a complex social network in which people, goods and information were exchanged, but as islands they were detached from the society around them, except at those moments when they opened their gates to devour new victims. Alexander Randall used the term 'archipelago' to characterize the shared, self-contained culture of American overseas military bases and their social isolation from their host communities (Randall 1986: 61). Following Solzhenitsyn, many authors have given the term a somewhat sinister cast. Michel Foucault developed Solzhenitsyn's concept into the idea of a 'carceral archipelago' of state institutions whose purpose was to discipline and punish society and which were able to do so precisely because they were dispersed through society, much like an archipelago (Foucault 1982: 302). Others have described as archipelagos so-called 'gated communities', middle-class residential districts in large cities that protect themselves from urban criminality by building walls and setting up controlled entry points to whole clusters of houses (Rodgers 2004: 114).

---

4 'Indonesia "finds" 500 islands', *BBC News*, 20 February 2003, available at http://news.bbc.co.uk/2/hi/asia-pacific/2784461.stm, accessed 9 November 2008. For further discussion, see Hopper (1978: 35–8), Cribb (2000: 1–10) and Emmerson (2005: 11).

In Indonesia itself, the New Order's system of camps for suspected communist detainees following the 1965 coup also lends itself to an archipelagic metaphor (Cribb 2000: 171). More recently, Tom Boellstorff has identified a 'gay archipelago' characterized by what he calls 'islands of difference' — pockets of gay and lesbian culture embedded within but socially isolated from the majority heterosexual culture (Boellstorff 2005: 7). In this usage, 'archipelago' as a social term starts to approach the term 'diaspora' in meaning, differing only in that diasporas are generally presumed to relate to some (perhaps imagined) homeland, whereas an archipelago is unambiguously home to its people.[5]

In short, there is an archipelagic dynamic, in which a sense of unity emerges despite dispersal and distance, that appears to operate in metaphorical archipelagos as much as in real ones. Indonesia's archipelagic character creates special problems, and a few special opportunities, for the Indonesian state.

## THE SEA IN INDONESIAN LIFE

Very different from the task of managing islands separated by the sea is the task of managing the sea itself. The vast area of sea that lies within and around the Indonesian archipelago is crucial to the Indonesian state for many reasons. The sea is an avenue of transport and communication. It connects Indonesia's islands with each other and it connects Indonesia with the rest of the world, carrying the imports and exports that sustain the economy. There are more than 750,000 dockings per year at Indonesian ports and those ports load more than 300 million tonnes of cargo per year.[6] About 14 million Indonesians travel by sea each year to seek work, carry on business or visit family members.[7] Marine transport is provided by 1,156 registered shipping companies and approximately 10,000 vessels (ESCAP 1999: 30).

---

5   For a study that comes close to conflating archipelagos and diasporas, see Goldstein (2000).

6   See Dick (1987); Rutz and Coull (1996); Statistics Indonesia, 'Inter island and international cargo loading and unloading, Indonesia 1988–2005', available at   http://www.bps.go.id/sector/transpor/sea/yearly/table4.shtml,   and 'Number of ship's call at commercial and non commercial sea port Indonesia, 1995–2005', available at http://www.bps.go.id/sector/transpor/sea/yearly/table3.shtml, both accessed 9 November 2008.

7   Figures record 14,737,000 passengers boarding commercial and non-commercial shipping in Indonesia in 2005, but only 13,664,000 disembarking; see Statistics Indonesia, 'Number of ship's passenger at commercial and non commercial sea port, Indonesia 1995–2005', available at http://www.bps.go.id/sector/transpor/sea/yearly/table5.shtml, accessed 9 December 2008.

The sea is also a source of vulnerability, a potential highway for ene-
mies and a conduit for unwanted people and goods. Colonial rule arrived
in the archipelago aboard the ships of foreign powers. Indonesia is not
a major final destination for illegal immigrants, but there is a significant
flow of illegal entrants from the Middle East who either intend to stay
or end up staying (OECD 2003: 254; Hunter 2006). In 2007, around 1,000
Burmese seafarers were stranded in Tual after being discharged without
papers by the Thai-flagged ships on which they were formerly employed
(Higginbottom 2007: 19). The transit of illegal immigrants through Indonesia to Malaysia and Australia has been a significant problem in relations with those two countries. Large numbers of Indonesians also cross
illegally into neighbouring countries, most often for work (Hugo 1993;
Ford 2006). In addition, there is a vast number of illegal border crossings
which are not expected to lead to permanent settlement. They include
border crossings by small-scale traders, sometimes in illegal goods, and
fishermen.

The sea is also an important economic resource. A wide range of
marine life is harvested for human consumption – not just fish (including
sharks) and shrimp, but also turtles, tripang and shellfish (Bentley 1996;
Williams 2007: 40–42). It has been estimated that fish comprise 60 per
cent of the protein consumed by Indonesians (Tomascik et al. 1997: 1,185;
Williams 2007: 43–4). Marine life collected for commercial purposes also
includes trochus, clams and other seashells, aquarium fish, coral and seaweed (Bentley 1999; Fougères 2008). The vast bulk of harvested marine
life is caught wild. The Indonesian seas include some of the world's richest fishing grounds and Indonesia is the world's fourth largest producer
of fish after China, Peru and India. There are enormous practical obstacles to measuring the size of the annual catch of any marine life, but plau-
sible estimates of the current annual catch range from 3.7 to 7.7 million
tonnes, compared with 2.8 million tonnes in 1994 (Delfs 2006).

Fishing has traditionally provided incomes for millions of people in
coastal villages across the archipelago. Although capture fishing has been
the most important source of employment in the marine sector, the farming of shrimp takes place on a vast scale in the coastal regions of western Indonesia, and the cultivation of milkfish (*bandeng*, *Chanos chanos*)
is increasingly important. The government began to promote shrimp
farming in the mid-1970s and to develop intensive farming techniques
from the mid-1980s. In 1995 a catastrophic viral disease struck most of
the shrimp farms, reducing production by 90 per cent, but production
eventually resumed in southern Sumatra. Although an Indonesian firm,
PT Central Proteinaprima, is the world's largest cultivated shrimp producer, the total area of shrimp farms declined from 231,460 hectares in
1989 to 132,800 hectares in 2005. A large part of the production (279,543

tonnes in 2005) is exported, and shrimp has become an important source of foreign exchange. Seaweed is also cultivated sporadically, and aquaculture is reported to employ about 2.5 million people in Indonesia.[8]

In recent decades the sea has become increasingly important as a source of minerals. The first mineral to be extracted from the sea was salt, which was harvested from early times on the beaches of many islands in the archipelago (Knaap and Nagtegaal 1991; Butcher 1996). The offshore dredging of tin began in the 1890s (Moolhuijzen 1972). Offshore extraction of oil began in 1971 with Arco's opening of the Cinta and Arjuna fields off West Java. This area remains the richest offshore oil region in Indonesia, though production is now in decline. Also important from the 1970s were the fields off the coast of Southeast Sumatra. More recently, both shallow and deep-sea oilfields have been discovered in the Makassar Strait (Ooi Jin-bee 1982: 12; Tangsubkul 1982: 82; Barnes 1995: 79).

Extraction of natural gas off the coast of West Java began in the early 1970s, and soon expanded to the eastern Java Sea and the seas around the Natuna Islands. The geomorphology of eastern Indonesia led many geologists to doubt that significant oil or gas reserves would be found there, but Arco discovered a huge offshore gas field at Tangguh in Bintuni Bay in 1997 and there are reports of major offshore fields in Aceh. In all, Indonesia is estimated to have around 2.8 trillion cubic metres of proven offshore natural gas reserves.[9] Increasing attention is also being given to other seabed minerals. Early interest in polymetallic nodules (formerly known as manganese nodules) has been followed by attention to hydrothermal sulphides containing gold, silver, copper, lead and zinc, which have been reported from the Sulawesi Sea north of Manado, and to cobalt-rich ferro-manganese crusts.[10] Indonesia is also estimated to have on its seafloor 24.3 trillion cubic metres of methane hydrate, regarded by some experts as a major future source of energy to supplement oil and gas.

In recent times, sand and coral have been mined on an increasing scale for building and construction purposes (Bentley 1998). Until 2002, there

---

8  See Siregar (1999); ADB (2006); USDA Foreign Agricultural Service (2008); 'Indonesia's CP Prima to spend US$242 mln on shrimp production', *Panorama Acuicola*, 7 December 2007, available at http://www.panoramaacuicola.com/ noticia.php?art_clave=4200, accessed 9 December 2008. On seaweed cultivation, see Backhaus and Wälty (1995).

9  'Official energy statistics from the U.S. government: Indonesia', available at http://www.eia.doe.gov/emeu/cabs/Indonesia/NaturalGas.html, accessed 3 March 2009.

10  'The dawn of deep ocean mining', *ScienceDaily*, 21 February 2006, available at http://www.sciencedaily.com/releases/2006/02/060221090149.htm, accessed 5 November 2008.

was a significant export of sand mined from the islands and seabeds of the Riau Archipelago to Singapore, where it was used for land reclamation purposes. This trade was banned in 2002 because of the environmental damage that mining was causing in Riau (Simamora 2002). The mining of sand to replenish tourist beaches has also been reported from Bali (Atmodjo 2008).

The sea can be a hazard to human existence. The perils of shipwreck from storm or reef, or simply of individual misadventure, loom larger at sea than do comparable dangers on land. Pollution may not only damage fish stocks but also cause human illness when it passes into the human population through fish. Indonesia is largely outside the tropical cyclone zone, so its coastal regions are generally not subject to the intense winds and storm surges that repeatedly cause enormous loss of life in neighbouring countries, but local storms at sea can be catastrophic for fishermen and ferries. Indonesia's location in a zone of geological instability means that many of its coastal regions are vulnerable to tsunamis. The 2004 tsunami that devastated Aceh and claimed an estimated 128,000 lives was the worst of a long series of such events associated with earthquakes and volcanic eruptions (Simkin and Fiske 1983). We are just beginning, moreover, to understand the role of the seas in climatic patterns. The consequences of the El Niño Southern Oscillation for rainfall in Indonesia have been known to scientists for nearly a century, but have only become part of broader public knowledge in the last decade or so (Nicholls 1992: 167).

The sea also has a special, if ambiguous, place in perceptions of Indonesian identity. Claiming a special affection for the sea, or an attachment to it, contemporary Indonesians often use a poetic term for the country, *tanah air* (land and water), to conjure up a sense of the unity of land and sea in the Indonesian national imagination (see, for example, Moertopo 1978: 73-5). Yet the expression appears to be a relatively recent one — it first appears not in the classical texts of Malay civilization but rather in the 1842 *Hikayat Abdullah* [The Tale of Abdullah], a work that is distinctively modern in vocabulary and world-view.[11] Moreover, the most celebrated appearance of the term *tanah air* in nationalist literature — in Muhammad Yamin's poem *Tanah Air* — refers to the fresh water that nourishes the land, not to the sea that surrounds it. Another commonly cited poetic term, *nusantara* (islands in between), draws attention to Indonesia's archipelagic character, but does not focus on the sea itself (Kusumaatmadja 1982). The best known sign of cultural attachment between Indonesians and the sea is the widespread belief in the legend

---

11   See the Malay Concordance Project, available at http://mcp.anu.edu.au/N/ Abd.H_bib.html, accessed 6 November 2008.

of Nyai Loro Kidul, the goddess of the southern seas who is the spiritual consort of the rulers of Java according to one set of legends. These stories, however, suggest that the sea is a counterpoint to the human world of the land, a place where humans do not really belong rather than an integral part of human existence.[12]

Emotional connection with the sea, like attachment to land, is a real cultural phenomenon that is notoriously difficult to measure and seriously prone to platitude and exaggeration. It is partly a consequence of a prosaic sense of dependence on the sea for livelihood, especially in fishing communities. In more complex ways it has to do with the sense of having special knowledge about the sea — an ability to 'read' the sea, for instance, or to identify an underwater hazard such as a reef, or to foresee a change of wind or the approach of a storm — that enhances the prospects for survival in a dangerous environment.[13] Such an emotional connection arises from and is expressed in the belief that the sea itself has a special spiritual character or is the home of spiritual beings, such as the 'sea wives' of pearl divers in the Aru Islands (Spyer 1997). It may even relate to the use of the sea for leisure activities. Although there is a substantial literature on the role of entertainment in Indonesian cultures, little attention has been paid to the place of leisure in general in modern Indonesia.[14] Anecdotal evidence suggests that Indonesia has only a meagre culture of maritime leisure — surfing is mainly the preserve of foreign visitors, as is recreational sea-fishing, diving and yachting.[15] A cultural history of Indonesian imaginings of the sea is yet to be written.

It is safe to say, however, that attachment to the sea is unevenly spread among Indonesia's many peoples. A few communities — the Orang Suku Laut (Sea People) of western Indonesia and the Bajau of eastern Indonesia — traditionally lived in intimate association with the sea, not only drawing their livelihood from it but living on it aboard their vessels (Chou 2003; Lowe 2003). For others — peasant farmers in Java, plantation workers in Sumatra, mountain dwellers throughout the archipelago — the sea is a remote and alien place, with no more than a minor part in their national imagining. There are strong signs, too, that the Orang Suku Laut and Bajau are losing their former close association with the sea. The

---

12  On Nyai Loro Kidul, see Jordaan (1984) and Wessing (1988: 56-7).

13  On the creation of attachment to land (and, by implication, the sea) that arises from surviving the hazards it presents, see Narangoa and Cribb (2003).

14  See, however, the important work by van Leeuwen (1997, 2005).

15  The late President Soeharto was fond of big-game fishing, but this enthusiasm did not spread widely within the Indonesian elite. On Indonesia as a destination for foreign surfers, see Lueras et al. (2002); for a glimpse of the early history of recreational use of the sea in Indonesia, see Rinkes, van Zalinge and de Roever (1927: 209-40).

need to educate their children is a powerful incentive for them to move onshore, and local governments also encourage it as a means of social betterment (Hajramurni 2008). As land-based communications improve, too, many coastal communities are becoming gradually less oriented to the sea in their day-to-day lives.

## MANAGING THE SEAS

Governing the maritime realm has always posed distinct challenges to governments, different from those of ruling the land. The physical fluidity of the sea has three far-reaching consequences for governance. First, it is always difficult (and mostly impossible) to mark any clear border on the sea or to place an installation at any fixed point on the sea's surface. This absence of fixity greatly complicates any attempt to establish or maintain a governmental presence at sea or to make the extent of a government's claims clear to visitors. Second, the sea is dangerous. Most of the hazards found on land exist in one form or another at sea as well, but hardly anywhere on land is the possibility of death as close and as constant as it is at sea. Third, the sea is multi-layered. Governance regimes on land normally make a distinction between air column, surface and subsoil. Governance at sea, in contrast, must take account of air column, surface, water column, seabed and subsoil, each of them different realms. This multi-dimensionality means that different governance challenges may exist at different levels at a single spot on the map.

These difficulties of governance mean that most activities at sea require a higher level of technology and skill than do comparable activities on land. For this reason, changes in the mastery of technology have historically been of huge importance in determining access to and control of the sea. Developments in European shipping design and navigational technology brought European ships to Indonesia from the sixteenth century, leading to a profound transformation of the archipelago (Glete 2000: 87–8). The ability to mount cannon on ships cemented Western supremacy. Later advances in the technology of navigation and mapping have produced dramatic changes in the capacity to locate and harvest fish and minerals at sea.

Although Indonesians have a long history of seafaring prowess, they have been relatively weak in mastering new technologies that would enable the government to control its seas. Lack of capital and of state revenue is one source of the problem. Although Indonesia has enjoyed substantial income for many years from the export of natural resources, there has always been significant internal competition for that income, and the relatively high costs of marine technology have been a constant disadvantage in the struggle for budget allocations. Indonesia's rulers, moreover, have a long tradition of making a virtue of the lack of access to technology by emphasizing political solutions over technological ones. During the independence struggle against the Dutch, leaders like Sukarno consciously adopted a strategy of embedding nationalist consciousness among the people, forecasting that Dutch power would fall away if Indonesians could be persuaded out of their deference to colonial authority. In stark contrast, for instance, to Leninist revolutionaries, the Indonesian nationalists paid little attention to the technologies of revolution that might have brought them to power (Legge 1972: 102–3, 115).[16] The same reliance on popular will was evident when, faced with superior Dutch arms, Indonesia's military leaders developed the defence doctrine known as Hankamrata (Pertahanan Rakyat Semesta), or Total People's Defence, under which what was said to be an integral relationship between the people and the armed forces compensated for a lack of sophisticated weaponry (Cribb 2001).

Even under the New Order (1966–98), when significant sums were invested in infrastructure and technology, the dominant mood in cabinet tended to be hostile to interventionist, technological fixes. The so-called 'technocrats' who dominated New Order policy making for at least two decades from 1966 took a neoclassical approach that stressed macroeconomic settings and avoided direct economic interventions. Only with the rise in influence of the technology minister, B.J. Habibie, in the second half of the 1980s did the so-called 'technologs' win a serious say in government, precipitating ambitious spending on projects whose attraction lay in attempts to master advanced technology (Hill 1995). Even then, Habibie was in constant conflict with those who favoured a more political and less technological approach to government.

The consequent lack of mastery of maritime technology has been reflected in several enduring management problems at sea. Indonesia has been generally weak in combating sea-borne criminal activities. Smuggling was widespread during the colonial era (Tagliacozzo 2005) and continued to be important after Indonesia became independent. During the war of independence, the Indonesian government itself sponsored clandestine trade to obtain funds for the struggle against the Dutch (Homan 1983; Cribb 1988). The problem was most acute in the 1950s and 1960s, when central government policies were hugely disadvantageous to exporters, many of whom were located close to alternative markets

---

16   Indeed, the communist leader D.N. Aidit evidently relied in much the same way on popular spirit and paid the same scant attention to technique in his attempt to shift political power towards the Communist Party in October 1965 (Roosa 2006).

across the maritime border. The scale of smuggling may have diminished during the New Order but it remained a serious problem (Simkin 1970). These days it includes, on the one hand, the import of drugs, of items that undercut Indonesia's own industries (ranging from processed steel to used clothing) and of contraband that circumvents Indonesian tariffs (such as alcohol and tobacco), and, on the other hand, the export of cash, endangered wildlife species, antiquities and subsidized fuel and fertilizer. The problem remains so serious that the government recently considered closing the Sumatran port of Bagan Siapi-Api to foreign trade because of widespread smuggling and underinvoicing (Dameria 2008). Djoko Sumaryono (Chapter 8) notes both the difficulty of coordinating the different government agencies responsible for maritime affairs and the recent achievements of Indonesia's Maritime Security Coordinating Board (Badan Koordinasi Keamanan Laut, or Bakorkamla).

Piracy, especially in the Malacca Strait, is also a serious problem, although Sam Bateman (Chapter 7) notes significant improvement in recent times. In the aftermath of Soeharto's fall in 1998, there was a weakening of surveillance in maritime areas, with the consequence that there were 113 piracy incidents — more than a third of the world's reported pirate attacks — in Indonesian waters in 1999. In 2002, Indonesia commenced joint patrols with India in the Andaman Sea approaches to the Malacca Strait, and in 2004 Indonesia, Malaysia and Singapore began joint patrols in the strait itself. In 2005, the United States offered to donate 40 coast-guard ships, old but still serviceable, to the Indonesian authorities to help them patrol the Malacca Strait. Nonetheless, Indonesian authorities acknowledge that they still have difficulty marshalling the capacity to patrol the region effectively. In line with the government's traditional preference to seek a political rather than technological solution, counterpiracy measures have included efforts to improve the standard of living in coastal districts (*kabupaten*) along the Malacca Strait (Rokan Hilir, Bengkalis, Siak, Palawan, Indragiri Ilir, Karimun), as well as other regions near major sea lanes. The expectation is that improved economic conditions will diminish the attraction of piracy.[17]

Indonesia faces a similarly perennial problem with illegal fishing. During the 1970s, serious concerns began to arise over the possibility of overfishing in Indonesian waters, leading to the banning of trawling in 1980, except in the Arafura Sea. This measure was followed by others to limit catches and restrict the access of foreign vessels to Indonesian

---

17  See Dillon (2000); Eklöf (2006); Ho (2006: 565–6); Permal (2006); 'India, Indonesia hold joint patrol', *Jakarta Post*, 13 October 2008, available at http://www.thejakartapost.com/news/2008/10/13/india-indonesia-hold-joint-patrol.html, accessed 10 December 2008.

waters. Nevertheless, huge volumes of fish, shellfish and other marine life are caught illegally within the archipelago and substantial amounts leave in the holds of foreign fishing vessels operating without permits, or with permits that have been transferred illegally from Indonesian licence holders to foreign fishers (Agoes 2005; Williams 2007). As with piracy, the government's instinctive response to illegal fishing has been to emphasize political and social solutions, rather than simple surveillance and punishment, recognizing the poor social conditions in fishing communities that lead to the infringement of regulations.

Indonesia also has a relatively weak record in developing regulatory regimes for activities at sea. As Sarah Waddell indicates in Chapter 11 of this volume, the overall legal regime governing Indonesian fisheries is relatively underdeveloped, despite the passing of a new Fisheries Law (Law No. 31/2004) in 2004 (Patlis 2007).[18] Indonesian waters encompass a vast range of ecological zones, including mangrove swamps, coral reefs, seagrass beds, open ocean and deep sea trenches. The conservation of these resources has been marked by sophisticated rhetoric in Jakarta and deficient delivery of results in the field (Arnscheidt 2009). The issue of marine pollution remains seriously underaddressed. As the country industrialized during the later decades of the Soeharto era, scant attention was paid to the dumping of industrial waste in rivers, which in turn polluted the seas (Cribb 1990: 1,124–5; Djalal 2000). The consequence was repeated instances of damage to coastal capture fisheries and aquaculture. The prosecution of offenders was infrequent.

Indonesia's lack of mastery of modern technology has also been apparent in its handling of *legal* economic activities at sea. In the legal fishing industry, Indonesian enterprises are generally outclassed by foreign ventures. The Indonesian section of the industry is dominated by small, artisanal fishers, while the high-tech sector is dominated by foreign or joint venture firms. Indonesia has maintained somewhat closer control of the extraction of oil, gas and other minerals from the sea but it remains largely dependent on foreign-owned technology and entrepreneurship in these sectors.

Marine safety is another major issue. As Erwin Rosmali demonstrates in Chapter 9 of this volume, Indonesia has witnessed repeated ferry disasters and accompanying loss of life. Several elements contribute to the problem of safety in Indonesian waters. Many vessels are old, and not all captains, navigators and other seafarers are properly trained. Many vessels lack basic navigational equipment. The checking of safety equipment is often rudimentary and vessels are often overloaded. Indonesia's search and rescue capacity is limited. As a result, accidents caused by bad

---

18   On the history of fishing in Indonesia, see Butcher (2004).

weather or faulty equipment can lead to extensive loss of life. Fishing and smaller commuter vessels are also regularly lost at sea. Nor does Indonesia have a strong record in managing the interests of the approximately 72,000 Indonesians who are reported to work as seafarers, 60 per cent of them on Indonesian ships and the remainder on foreign-registered vessels (Sijabat 2002). Indonesia is one of the world's largest sources of seafarers for international shipping. Few of them speak English, however, or have the training to work on larger vessels, and a recent International Labour Organization (ILO) report suggests that conditions in the industry are poor (ILO 2002). Many seafarers do not receive the minimum wage,[19] and work in conditions that breach standard safety requirements. Many have to pay up-front to get their jobs and are required to sign two contracts, one for the record that meets ILO standards and another that reflects actual conditions of work. The Indonesian Seafarers Union (Kesatuan Pelaut Indonesia, or KPI), established by the Directorate General of Marine Transport, has not been an effective negotiator with shipowners and is not recognized by the ILO. The extensive use of flags of convenience by vessels operating in Indonesian waters has added to the difficulty of monitoring conditions.[20]

Indonesia has been slow to develop an efficient network of interisland passenger and cargo services. When the New Order came to power in 1966, there were no formal passenger services between Indonesia's islands; passengers had no choice but to travel by cargo ship, mostly on deck. Under the Perintis project, the New Order government agreed to subsidize specific regular services, but the initiative began poorly when the first ferry introduced under this venture, the state-owned, second-hand, roll-on/roll-off *Tampomas II*, sank in 1981 with the loss of 580 lives. Between 1983 and 1987, the government purchased six new vessels from Germany and with them provided regular passenger services between the major ports for the first time since 1957. Nine more ships entered service between 1987 and 1994 (Rutz 1987: 491; Rutz and Coull 1996: 278;

---

19  A 1996 survey indicated that Russian, Ukrainian, Croatian and Indonesian seafarers received the lowest rates of pay in the industry, regardless of the country of registration of the ships on which they were working (DeSombre 2006: 137).

20  'Seafarers demand better payment, welfare', *Jakarta Post*, 30 July 2002, available at http://www.thejakartapost.com/news/2002/07/30/seafarers-demand-better-payment-welfare.html; 'Uphold safety management in RI waters: seafarers', *Jakarta Post*, 9 April 2007, available at http://www.the jakartapost.com/news/2007/04/09/uphold-safety-management-ri-waters-seafarers.html; 'Shipping woes due to govt failings', *Jakarta Post*, 5 November 2008, available at http://www.thejakartapost.com/news/2008/11/05/shipping-woes-due-govt-failings.html, all accessed 9 November 2008.

Dick 2008). Until 1985 the interisland shipping system was highly regulated, with access to particular routes based largely on a licensing system, but reforms carried out between 1985 and 1988 removed many of these restrictions (Dick and Forbes 1992; ESCAP 1999). As we see from David Ray's chapter in this book, Indonesia has also lagged in managing its port facilities. Indonesian ports are notorious for their slow turnaround times, lack of facilities and general poor management, a situation that represents an unnecessary tax on interisland trade.

Still less impressive is Indonesia's management of its maritime heritage. Thanks to the archipelago's long history of maritime commerce, the Indonesian seas are littered with shipwrecks dating back at least two millennia. These wrecks are potentially a source of great insight into the early commercial and cultural history of the region (Flecker 2001), but Indonesia has not given high priority to protecting its marine archaeological heritage. Shipwrecks and other marine heritage sites do not come under the authority of the Ministry of Culture and Tourism, which is responsible for archaeological sites on land, but rather under the Ministry of Marine Affairs and Fisheries, which has little institutional reason to be interested in heritage. When a tenth-century shipwreck containing rubies, glass and thousands of pieces of pottery was discovered by fishermen off the northern coast of West Java in 2000, the response of the ministry was to retain 10 per cent of the cache and auction off the rest. 'It has more economic value than historical value, there is no need to take it for our heritage museums', said a ministry spokesperson (Cuno 2008: 46). The growing sophistication of underwater scanning technology has exposed increasing numbers of shipwrecks to illegal plunder for the international antiquities market, but it is still the case that most new shipwreck sites are discovered by trawlers and diving fishermen (Flecker 2002).

Indonesia's naval defences have not been tested in conflict since the 1960s. Indonesia's naval encounter with the Dutch off Vlakke Hoek (Tanjung Namaripi) in 1962 during the confrontation over West New Guinea (now Papua) ended in defeat, although the threat of naval action on a larger scale was one element in the eventual Dutch decision to give up control of the territory (Slot and Hendriks 2002). Throughout the period since independence, the Hankamrata doctrine has meant that Indonesian defence planning has given little attention to the navy as a first line of defence against foreign intrusion; the consequence has been small budgets and a relatively low general standing for the navy within the armed forces (Suryohadiprojo 1973; Dupont 1996).[21] Indonesia's most

---

21  On the history of the Indonesian navy, see Cahyono (1992) and Leirissa (2006).

effective innovation in marine surveillance, in fact, has been a political rather than a technological one: the government has recruited fishermen as part of its national security system under an arrangement known as *sistem pengawasan masyarakat* (*siswasmas*), or community surveillance system (Patlis 2007: 219). Pressure on Indonesia to enhance its technological capacity for surveillance has increased in recent times with the emerging fear of a possible terrorist attack in Southeast Asian waters. The principal fear has been that a hijacked vessel might arrive in Singapore via Indonesian waters as a floating explosive device, causing serious loss of life, extensive damage to infrastructure and a loss of confidence in the main container route through the Malacca Strait (Luft and Korin 2004; Sukma 2005; Ho 2006: 563–5).

If Indonesia has performed poorly in mastering the technology of managing the seas, it has also been unimpressive in its management of technology that might help to transcend Indonesia's archipelagic character. Except in the Riau Islands, where the then technology minister, B.J. Habibie, took the initiative to build a chain of six bridges connecting the islands of Batam, Rempang and Galang (Chou and Wee 2002: 351–2), Indonesia has not followed the example of other sea-bound countries, such as Denmark, Norway and Japan, in constructing bridges to connect islands. Although there has been a proposal for a bridge between Java and Madura, and between Java and Bali, neither has moved beyond the speculative stage (Leinbach and Ulack 1999: 221; Pringle 2004: 220–21).

Although the record of Indonesia's technological management of the sea is a somewhat dismaying account of indifferent performance, the Indonesian inclination to seek effective political solutions has generated an impressive record in developing a political regime to govern the sea. The challenge in creating a political regime in maritime areas lies not in making laws that are consistent with the national interest, social rights and natural justice, but rather in producing laws that reflect both the specific character of the sea itself and the current capacity of states to govern it. Laws that cannot be implemented serve little purpose, as was shown centuries ago by the Treaties of Tordesillas (1494) and Zaragossa (1529), in which Spain and Portugal agreed to divide the world, including the oceans, between them. Although this division was endorsed by the Pope, it quickly became a dead letter because the Iberian powers were unable to enforce their claims against the intrusion of the northern Europeans.

Indonesia's contribution to the Law of the Sea since the 1950s has been strong. The articles in this volume by John G. Butcher (Chapter 2), Arif Havas Oegroseno (Chapter 3), Hasjim Djalal (Chapter 4) and I Made Andi Arsana and Clive Schofield (Chapter 5) reflect the fact that, for more than five decades, Indonesia has been at the forefront in devising new laws to reflect changing governmental capacity at sea. Most important of all has been Indonesia's development of the archipelagic state as a concept in international law. The consequence has been to create a new category of jurisdiction, archipelagic waters, which are recognized as part of the maritime territory of some two dozen archipelagic states around the world. The establishment of the concept of archipelagic waters has in turn required – in the case of Indonesia and the Philippines – the development of a regime for transarchipelagic passage known as archipelagic sea lanes.

In negotiating its maritime boundaries with Australia, Indonesia pioneered innovative means of recognizing the complexity of maritime resource exploitation. Under the Timor Gap Treaty of 1989, Indonesia and Australia shelved what had proven to be irreconcilable differences over the principle on which the maritime zone between Australia and what was then the Indonesian province of East Timor should be divided. Instead, they created a 'zone of cooperation', divided into three separate areas of exploitation, with each country having a different share of the responsibilities for each area (Prescott 1993c).[22] Then, in 1997, Indonesia and Australia signed a further treaty demarcating the entire EEZ boundary between the two countries. The innovative feature of this treaty was that it created separate, overlapping EEZ regimes for the water column (including fishing rights, which went to Indonesia) and the seabed (including mining rights, which went to Australia) (Prescott 1993a, 1993b, 1997).

One of the challenges for contemporary Indonesia is to decide whether and how to extend the imaginative approach to marine territoriality to the issue of marine tenure. In Western law, which forms the principal basis of Indonesian law, private ownership of the sea is not possible, and there is no easy way to recognize the traditional rights of individuals over areas of sea. Where such recognition exists in practice, it most commonly takes the form of licences granted by the state, on its own authority, to individuals or groups whom it judges to be traditional users. Customary marine tenure – that is, the public recognition of traditional ownership of marine resources – is common in parts of eastern Indonesia, despite evidence that it is a relatively recent construction (Osseweijer 2001: 111–15; Bubandt 2005), but it receives no support in Indonesian law. The Indonesian constitution, moreover, is well known for explicitly subordinating private ownership of land to the broader national interest. Article 33(3) states succinctly: 'The land, the waters and the natural riches contained therein shall be controlled by the State and exploited to the greatest benefit of the people'. The legal environment for recognition of

---

22   This treaty became a dead letter in 1999 with the independence of the province as Timor-Leste.

customary marine tenure is thus unpromising. James J. Fox (Chapter 12) draws attention to the difficulties the Australian government has faced in regulating 'traditional' uses of marine resources in regions that are now within Australia's jurisdiction. An ironic consequence is that maritime and coastal communities such as the Orang Suku Laut have sought first of all to establish a claim to land. Cynthia Chou quotes a member of this community, which is famous above all for its identity with life at sea, speaking of territory purely in terms of land:

> For us, we have been on this land since we were young children. This is *tanah kami* (our land) and not the *tanah* of others. ... [O]ur father, mother and siblings are all buried here. We do not want to disturb them. ... We do not want to move to another island. We just want to live and die in this one island, which is ours (Chou 1997: 616).

Within Indonesia, however, conservation groups have been developing creative measures to combine marine conservation with traditional exploitation of sea resources (Satria, Sano and Shima 2006). There is some prospect that this approach will lead to significant changes in maritime environmental management. Rili Djohani (Chapter 10) suggests that the Indonesian government performs best in environmental protection (as in defence) when it emphasizes the political rather than technical aspects of management.

Indonesia is at a vital juncture in the management of its seas. Partly for historical reasons, partly thanks to its vigorous contributions to the framing of the Law of the Sea, Indonesia is recognized as possessing 5.8 million square kilometres of sea, and may soon come to possess more if its proposal for an extended continental shelf is accepted. With this vast area come substantial opportunities — the seabed will be one of the great resource frontiers of the twenty-first century and new opportunities for productively managing the ecology of marine areas are constantly emerging. Yet with this opportunity come responsibilities and threats. As a contributor to the international order, Indonesia will be obliged to exercise more direct authority over the seas under its control — protecting them from degradation and ensuring the safety of those who traverse them, while ensuring its own security and economic interests. Success in bringing effective governance to this vast realm will be crucial in Indonesia's economic and social development.

## ACKNOWLEDGMENTS

We would like to thank Sandra Wilson and Michael Flecker for helpful comments on earlier drafts of this chapter.

# REFERENCES

ADB (Asian Development Bank) (2006), 'Expanding aquaculture production for Indonesia's poor fish farmers', ADB news release, Manila, 14 December, available at http://www.adb.org/Media/Articles/2006/11167-indonesian-aquaculture-development-projects/, accessed 9 December 2009.

Acciaoli, Greg (2001), '"Archipelagic culture" as an exclusionary government discourse in Indonesia', *Asia Pacific Journal of Anthropology*, 2(1): 1–23.

Agoes, Etty R. (2005), 'Adequacy of Indonesian laws and regulations to combat IUU fishing: an evaluation of the new law on fisheries', 27 April, available at http://www.illegal-fishing.info/uploads/Indonesia-law-on-fisheries-evaluation.pdf, accessed 10 December 2008.

Arnscheidt, Julia (2009), *'Debating' Nature Conservation: Policy, Law and Practice in Indonesia*, Leiden University Press, Leiden.

Atmodjo, Wasti (2008), 'Governor: stop sand excavation', *Jakarta Post*, 27 September, available at http://www.thejakartapost.com/news/2008/09/27/governor-stop-sand-excavation.html, accessed 9 December 2008.

Backhaus, Norman and Samuel Wälty (1995), 'Seetanganbau auf Bali: neues geld und neue verunsicherung durch globalisierung' [Seaweed cultivation in Bali: new money and new uncertainty from globalization], in S. Wälty and B. Werle (eds), *Kulturen und Raum: Theoretische Ansätze und Empirische Kulturforschung in Indonesien* [Cultures and Space: Theoretical Stages and Empirical Research into Culture in Indonesia], Chur Rüegger, Zürich, pp. 391–405.

Barnes, Philip (1995), *Indonesia: The Political Economy of Energy*, Oxford University Press, Oxford.

Bentley, Nokome (1996), 'An overview of the exploitation, trade and management of chondrichthyans in Indonesia', in H.K. Chen (ed.), *An Overview of Shark Trade in Selected Countries of Southeast Asia*, TRAFFIC Southeast Asia, Petaling Jaya, available at http://www.trophia.co.nz/documents/Sharks&RaysIndonesia.pdf, accessed 10 December 2008.

Bentley, Nokome (1998), 'An overview of the exploitation, trade and management of corals in Indonesia', *TRAFFIC Bulletin*, 17(2), available at http://www.trophia.co.nz/documents/CoralsIndonesia.pdf, accessed 10 December 2008.

Bentley, Nokome (1999), 'Fishing for solutions: can the live trade in wild gropers and wrasse from South East Asia be managed?', TRAFFIC Southeast Asia, Petaling Jaya, available at http://www.trophia.co.nz/documents/Fishing ForSolutions.pdf, accessed 10 December 2008.

Boellstorff, Tom (2005), *The Gay Archipelago: Sexuality and Nation in Indonesia*, Princeton University Press, Princeton NJ and Oxford.

Bubandt, Nils (2005), 'On the genealogy of *sasi*: transformations of an imagined tradition in eastern Indonesia', in T. Otto and P. Pedersen (eds), *Tradition and Agency Tracing Cultural Continuity and Invention*, Aarhus University Press, Aarhus, pp. 193–232.

Butcher, John G. (1996), 'The salt farm and fishing industry of Bagan Si Api Api', *Indonesia*, 62: 90–120.

Butcher, John G. (2004), *The Closing of the Frontier: A History of the Marine Fisheries of Southeast Asia c. 1850–2000*, KITLV Press, Leiden.

Cahyono, Agus (1992), *Sejarah Singkat TNI Angkatan Laut: 1945–1985* [A Brief History of the Indonesian Navy: 1945–1985], Subdirektorat Sejarah, Direktorat Perawatan Personil TNI AL, Jakarta.

Chou, Cynthia (1997), 'Contesting the tenure of territoriality: the Orang Suku Laut', *Bijdragen tot de Taal-, Land- en Volkenkunde*, 153(4): 605–29.

Chou, Cynthia (2003), *Indonesian Sea Nomads: Money, Magic and Fear of the Orang Suku Laut*, RoutledgeCurzon, London.

Chou, Cynthia and Vivienne Wee (2002), 'Tribality and globalization: the Orang Suku Laut and the "growth triangle" in a contested environment', in G. Benjamin and C. Chou (eds), *Tribal Communities in the Malay World: Historical, Cultural and Social Perspectives*, Institute of Southeast Asian Studies, Singapore, pp. 318–63.

Cribb, Robert (1988), 'Opium and the Indonesian revolution', *Modern Asian Studies*, 22(4): 701–22.

Cribb, Robert (1990), 'The politics of pollution control in Indonesia', *Asian Survey*, 30(12): 1,124–5.

Cribb, Robert (1999), 'Not the next Yugoslavia: prospects for the disintegration of Indonesia', *Australian Journal of International Affairs*, 53(2): 169–78.

Cribb, Robert (2000), *Historical Atlas of Indonesia*, Curzon, London.

Cribb, Robert (2001), 'Military strategy in the Indonesian revolution: Nasution's concept of "total people's war" in theory and practice', *War and Society*, 19(2): 143–54.

Cuno, James (2008), *Who Owns Antiquity? Museums and the Battle over our Ancient Heritage*, Princeton University Press, Princeton NJ.

Dameria, Olivia (2008), 'Govt may close down Bagan Siapi-Api port', *Jakarta Post*, 27 October, available at http://www.thejakartapost.com/news/2008/10/27/govt-may-close-down-bagan-siapiapi-port.html, accessed 9 November 2008.

Delfs, Robert (2006), 'Undoing the damage', *Asia Sentinel*, 9 August, available at http://www.asiasentinel.com/index.php?option=com_content&task=view&id=84&Itemid=175, accessed 9 December 2008.

DeSombre, Elizabeth R. (2006), *Flagging Standards: Globalization and Environmental, Safety, and Labor Regulations at Sea*, MIT Press, Cambridge MA.

Dick, H.W. (1987), *The Indonesian Interisland Shipping Industry: An Analysis of Competition and Regulation*, Institute of Southeast Asian Studies, Singapore.

Dick, H.W. (1996), 'The emergence of a national economy, 1808–1990s', in J.Th. Lindblad (ed.), *Historical Foundations of a National Economy in Indonesia, 1890s–1990s*, North-Holland, Amsterdam, pp. 21–51.

Dick, HW. (2008), 'The 2008 Shipping Law: deregulation or re-regulation?', *Bulletin of Indonesian Economic Studies*, 44(3): 383–406.

Dick, H.W and D. Forbes (1992), 'Transport and communications: a quiet revolution', in A. Booth (ed.), *The Oil Boom and After: Indonesian Economic Policy and Performance in the Soeharto Era*, Oxford University Press, Singapore, pp. 258–82.

Dillon, Dana Robert (2000), 'Piracy in Asia: a growing barrier to maritime trade', Heritage Foundation Backgrounder No. 1379, 22 June, available at http://www.heritage.org/research/asiaandthepacific/bg1379.cfm, accessed 10 December 2008.

Djalal, Hasjim (2000), 'The prevention of marine pollution in the Asia Pacific region: Law of the Sea Convention and its implications', *Indonesian Quarterly*, 28(2): 227–35.

Dupont, Alan (1996), 'Indonesian defence strategy and security: time for a rethink?', *Contemporary Southeast Asia*, 18(3): 275–97.

Eklöf, Stefan (2006), *Pirates in Paradise: A Modern History of Southeast Asia's Maritime Marauders*, NIAS, Copenhagen.

Emmerson, Donald K. (2005), 'What is Indonesia?', in John Bresnan (ed.), *Indonesia: The Great Transition*, Rowman & Littlefield, Lanham MD, pp. 7–73.

Encyclopaedie van Nederlandsch-Indië (1917), 'Bawean', *Encyclopaedie van Nederlandsch-Indië*, Volume 1, Martinus Nijhoff and E.J. Brill, The Hague and Leiden, pp. 211–12.

ESCAP (United Nations Economic and Social Commission for Asia and the Pacific) (1999), 'A pilot study on the alleviation of poverty in remote island communities in Indonesia', Report No. 2017, New York, available at http://www.unescap.org/publications/detail.asp?id=558, accessed 10 December 2009.

Flecker, Michael (2001), 'A ninth-century AD Arab or Indian shipwreck in Indonesia: first evidence for direct trade with China', *World Archaeology*, 32(3): 335–54.

Flecker, Michael (2002), 'The ethics, politics, and realities of maritime archaeology in Southeast Asia', *International Journal of Nautical Archaeology*, 31(1): 12–24.

Ford, Michele (2006), 'After Nunukan: the regulation of Indonesian migration to Malaysia', in A. Kaur and I. Metcalfe (eds), *Divided We Move: Mobility, Labour Migration and Border Controls in Asia*, Palgrave Macmillan, New York, pp. 228–47.

Foucault, Michel (1982), *Discipline and Punish*, Penguin, Harmondsworth.

Fougères, Dorian (2008), 'Old markets, new commodities: aquarian capitalism in Indonesia', in J. Nevins and N.L. Peluso (eds), *Taking Southeast Asia to Market: Commodities, Nature, and People in the Neoliberal Age*, Cornell University Press, Ithaca NY, pp. 161–75.

Glete, Jan (2000), *Warfare at Sea, 1500–1650: Maritime Conflicts and the Transformation of Europe*, Routledge, London.

Goldstein, Paul S. (2000), 'Communities without borders: the vertical archipelago and diaspora communities in the southern Andes', in M.A. Canuto and J. Yaeger (eds), *The Archaeology of Communities: A New World Perspective*, Routledge, New York, pp. 182–209.

Hajramurni, Andi (2008), 'Bajo people losing their identity', *Jakarta Post*, 22 October, available at http://www.thejakartapost.com/news/2008/10/22/bajo-people-losing-their-identity.html, accessed 9 November 2008.

Higginbottom, Katie (2007), 'Fishy business', *Seafarers' Bulletin*, 21: 18–19, available at http://www.itfseafarers.org/files/publications/3820/SB07En.pdf, accessed 9 November 2008.

Hill, Hal (1995), 'Indonesia's great leap forward? Technology development and policy issues', *Bulletin of Indonesian Economic Studies*, 31(2): 65–123.

Ho, Joshua H. (2006), 'The security of sea lanes in Southeast Asia', *Asian Survey*, 46(4): 558–74.

Homan, Gerlof D. (1983), 'American business interests in the Indonesian republic, 1946–1949', *Indonesia*, 35(April): 125–32.

Hopper, Richard H. (1978), 'Indonesia's 13,669 islands', *Indonesia Circle*, 17(November): 35–8.

Hugo, Graeme (1993), 'Indonesian labour migration to Malaysia: trends and policy implications', *Asian Journal of Social Science*, 21(1): 36–70.

Hunter, Cynthia (2006), 'People in between', *Inside Indonesia*, 88(October–December), available at http://insideindonesia.org/content/view/49/29, accessed 11 November 2008.

ILO (International Labour Organization) (2002), 'Report 1: report on an ILO investigation into the living and working conditions of seafarers', report for discussion at the Meeting of Experts on Working and Living Conditions of Seafarers on Board Ships in International Registers, Geneva, available at http://www-ilo-mirror.cornell.edu/public/english/dialogue/sector/tech meet/mewlcs02/mewlcs-r1.pdf.

Jordaan, Roy E. (1984), 'The mystery of Nyai Lara Kidul, goddess of the southern ocean', *Archipel*, 28: 99–116.

Knaap, Gerrit and Luc Nagtegaal (1991), 'A forgotten trade: salt in Southeast Asia 1670–1813', in R. Ptak and D. Rothermund (eds), *Emporia, Commodities and Entrepreneurs in Asian Maritime Trade, 15th to 18th centuries*, Steiner, Stuttgart, pp. 127–57.

Kusumaatmadja, Mochtar (1982), 'The concept of the Indonesian archipelago', *Indonesian Quarterly*, 10: 12–26.

Kwiatkowska, Barbara (1991), 'The archipelagic regime in practice in the Philippines and Indonesia: making or breaking international law?', *International Journal of Estuarine and Coastal Law*, 6(1): 1–32.

Legge, J.D. (1972), *Sukarno: A Political Biography*, Penguin, Harmondsworth.

Leinbach, Thomas R. and Richard Ulack (1999), *Southeast Asia: Diversity and Development*, Prentice Hall, Upper Saddle River NJ.

Leirissa, R.Z. (2006), *Rajawali Laut: 50 Tahun Penerbangan TNI Angkatan Laut* [Sea Eagles: 50 Years of the Air Arm of the Indonesian Navy], Panitia Penyusun Buku Sejarah Penerbangan Angkatan Laut Bekerja Sama dengan Red & White Publishing, Jakarta.

Lowe, Celia (2003), 'The magic of place: Sama at sea and on land in Sulawesi, Indonesia', *Bijdragen Tot de Taal-, Land- en Volkenkunde*, 159(1): 109–33.

Lueras, Leonard, Lorca Lueras, Jason Childs and Bernie Baker (2002), *Surfing Indonesia: A Search for the World's Most Perfect Waves*, Tuttle, Rutland VT and Tokyo.

Luft, Gal and Anne Korin (2004), 'Terrorism goes to sea', *Foreign Affairs*, 83(6): 61–71.

Moertopo, Ali (1978), *Strategi Kebudayaan* [Cultural Strategy], Yayasan Proklamasi, Centre for Strategic and International Studies, Jakarta.

Moolhuijzen, Alb. W.A. (1972), 'De geboorte van "offshore tin dredging in Indonesia"' [The birth of 'offshore tin dredging in Indonesia'], *Erts: Maandblad der Billiton Bedrijven*, 24(1): 6–13.

Muhammad, Chalid (2008), 'Ecological disasters and Indonesia's future threats', *Jakarta Post*, 22 December, available at http://www.thejakartapost.com/news/2008/12/22/ecological-disasters-and-indonesia039s-future-threats.html, accessed 18 March 2009.

Narangoa, Li and Robert Cribb (2003), 'Dynamics of land and identity in Pacific Asia: reflections on attachment to land', *International Journal of the Humanities*, 1: 1,093–102.

Nicholls, Neville (1992), 'Historical El Niño/Southern Oscillation variability in the Australasian region', in H.F. Diaz and V. Markgraf (eds), *El Niño: Historical and Palaeoclimatic Aspects of the Southern Oscillation*, Cambridge University Press, Cambridge.

OECD (Organisation for Economic Co-operation and Development) (2003), *Migration and the Labour Market in Asia 2002*, Paris.

Ooi Jin-bee (1982), *The Petroleum Resources of Indonesia*, Institute of Southeast Asian Studies, Singapore.

Osseweijer, Manon (2001), 'Taken at the flood: marine resource use and management in the Aru Islands (Maluku, eastern Indonesia)', PhD thesis, University of Leiden, Leiden.

Patlis, Jason (2007), 'Indonesia's new Fisheries Law: will it encourage sustainable management or exacerbate over-exploitation?' *Bulletin of Indonesian Economic Studies*, 43(2): 201–26.

Permal, Sumathy (2006), 'Indonesia's efforts in combating piracy and armed robbery in the Straits of Malacca', Maritime Institute of Malaysia Online, May, available at http://www.mima.gov.my/mima/htmls/papers/, accessed 10 December 2008.

Prescott, Victor (1993a), 'Australia–Indonesia (seabed boundaries)', in J.I. Charney and L.M. Alexander (eds), *International Maritime Boundaries*, Volume 2, Martinus Nijhoff, Dordrecht, pp. 1,195–205.

Prescott, Victor (1993b), 'Australia–Indonesia (Timor and Arafura seas)', in J.I. Charney and L.M. Alexander (eds), *International Maritime Boundaries*, Volume 2, Martinus Nijhoff, Dordrecht, pp. 1,207–18.

Prescott, Victor (1993c), 'Australia–Indonesia (Timor Gap)', in J.I. Charney and L.M. Alexander (eds), *International Maritime Boundaries*, Volume 2, Martinus Nijhoff, Dordrecht, pp. 1,245–328.

Prescott, Victor (1997), 'The completion of marine boundary delimitation between Australia and Indonesia', *Geopolitics*, 2(2): 132–49.

Pringle, Robert (2004), *A Short History of Bali, Indonesia's Hindu Realm*, Allen & Unwin, Sydney.

Randall, Alexander 5th (1986), 'The culture of United States military enclaves', in M.L. Foster and R.A. Rubinstein (eds), *Peace and War: Cross-cultural Perspectives*, Transaction Publishers, New Brunswick NJ, pp. 61–9.

Rinkes, D.A., N. van Zalinge and J.W. de Roever (1927), *Het Indische Boek der Zee* [The Indies Book of the Sea], Kolff, Weltevreden.

Rodgers, Dennis (2004), '"Disembedding" the city: crime, insecurity and spatial organization in Managua, Nicaragua', *Environment and Urbanization*, 16(2): 113–23.

Roosa, John (2006), *Pretext for Mass Murder: The September 30th Movement and Suharto's Coup d'État in Indonesia*, University of Wisconsin Press, Madison WI.

Rutz, W. (1987), 'Indonesia's sea transport system: a series of maps', *GeoJournal*, 14(4): 491–502.

Rutz, Werner O.A. and James R. Coull (1996), 'Inter-island passenger shipping in Indonesia: development of the system: present characteristics and future requirements', *Journal of Transport Geography*, 4(4): 275–86.

Satria, Arif, Masaaki Sano and Hidenori Shima (2006), 'Politics of marine conservation area in Indonesia: from a centralised to a decentralised system', *International Journal of Environment and Sustainable Development*, 5(3): 240–61.

Sijabat, Ridwan Max (2002), 'Seafarers demand better payment, welfare', *Jakarta Post*, 30 July, available at http://www.thejakartapost.com/news/2002/07/30/seafarers-demand-better-payment-welfare.html, accessed 25 March 2009.

Simamora, Adianto P. (2002), 'Singapore must help Indonesia curb illegal sand mining, activists say', *Jakarta Post*, 27 July, available at http://www.ecology asia.com/news-archives/2002/jul-02/thejakartapost.com_20020727.L01. htm, accessed 9 December 2008.

Simkin, C.G.F. (1970), 'Indonesia's unrecorded trade', *Bulletin of Indonesian Economic Studies*, 6(1): 17–44.

Simkin, Tom and Richard S. Fiske (eds) (1983), *Krakatau, 1883: The Volcanic Eruption and Its Effects*, Smithsonian Institution Press, Washington DC.

Siregar, Raja (1999), 'Indonesia to intensify shrimp farming', Third World Network, available at http://www.twnside.org.sg/title/1961.htm, accessed 9 December 2008.

Slot, Rob Bruins and Gerda Jansen Hendriks (2002), 'De slag bij Vlakke Hoek' [The battle of Vlakke Hoek], in A. van Liempt (ed.), *Andere Tijden*, Veen, Amsterdam, pp. 177–84.

Solzhenitsyn, Alexander (1974), *The Gulag Archipelago, 1918–1956: An Experiment in Literary Investigation*, Collins & Harvill Press, Sydney.

Spyer, Patricia (1997), 'The eroticism of debt: pearl divers, traders, and sea wives in the Aru Islands, eastern Indonesia', *American Ethnologist*, 24(3): 515–38.

Sukarno (1961), 'The birth of Pancasila', speech delivered on 1 June 1945, in Department of Foreign Affairs (ed.), *Toward Freedom and the Dignity of Man*, Jakarta.

Sukma, Rizal (2005), 'Indonesia's maritime security interests: terrorism and beyond', *Indonesian Quarterly*, 33(1): 2–8.

Suryohadiprojo, Sayidiman (1973), 'The territorial defence concept', *Indonesian Quarterly*, 1(4): 65–78.

Tagliacozzo, Eric (2005), *Secret Trades, Porous Borders: Smuggling and States along a Southeast Asian Frontier, 1865–1915*, Yale University Press, New Haven CT.

Tangsubkul, Phiphat (1982), *ASEAN and the Law of the Sea*, Institute of Southeast Asian Studies, Singapore.

Tomascik, Tomas, Annmarie Janice Mah, Anugerah Nontji and Mohammad Kasim Moosa (1997), *The Ecology of the Indonesian Seas*, Oxford University Press, Oxford.

USDA (United States Department of Agriculture) Foreign Agricultural Service, 'Indonesia fishery products shrimp report 2007, GAIN Report No. ID7024, available at http://www.fas.usda.gov/gainfiles/200707/146291660.pdf, accessed 9 December 2008.

van Leeuwen, Lizzy (1997), *Airconditioned Lifestyles: De Nieuwe Rijken in Jakarta* [Airconditioned Lifestyles: The New Rich in Jakarta], Het Spinhuis, Amsterdam.

van Leeuwen, Lizzy (2005), 'Lost in mall: an ethnography of middle class Jakarta in the 1990s', PhD thesis, University of Amsterdam, Amsterdam.

Wessing, Robert (1988), 'Spirits of the earth and spirits of the water: chthonic forces in the mountains of West Java', *Asian Folklore Studies*, 47(1): 43–61.

Williams, Meryl J. (2007), *Enmeshed: Australia and Southeast Asia's Fisheries*, Lowy Institute for International Policy, Sydney.

## 2 BECOMING AN ARCHIPELAGIC STATE: THE JUANDA DECLARATION OF 1957 AND THE 'STRUGGLE' TO GAIN INTERNATIONAL RECOGNITION OF THE ARCHIPELAGIC PRINCIPLE

*John G. Butcher*

One of the fundamental features of Indonesia is that it is an archipelagic state. Large-format maps of Indonesia usually show the straight base-lines that join the outermost points of the outermost islands, thereby enclosing within a single entity the thousands of islands that make up the country and all the 'archipelagic waters' within the baselines (see Map 2.1). These maps also show the territorial sea that, except in a very few areas, extends 12 nautical miles out from these baselines and the exclusive economic zone (EEZ) that in most areas extends 200 nautical miles out from these same baselines. As a consequence much of the territory of Indonesia is made up of water. More precisely, water makes up about 58 per cent of the total area of 4.5 million square kilometres over which the state asserts its sovereignty, namely the land and water within the baselines and the territorial sea extending out from the baselines.[1] The country's EEZ, within which the government exercises sovereign rights over all the living and non-living resources in the water and both on and under the seabed, gives the government limited jurisdiction over a further 5.4 million square kilometres. The result is a massive country.

---

1 These calculations are based on data in Forbes (1995: 2, 15).

*Map 2.1   Indonesia's archipelagic baselines and sea lane passages*

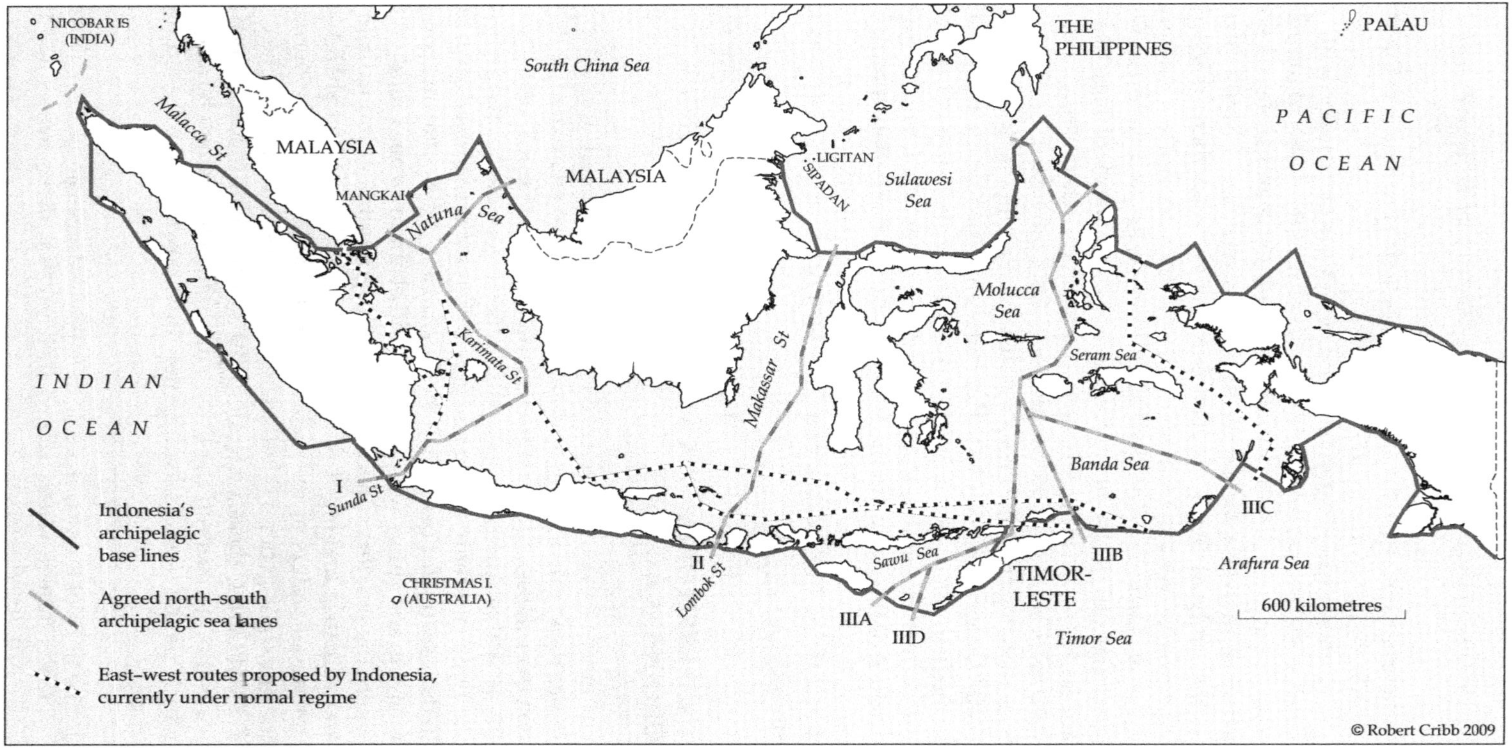

PALAU

THE
PHILIPPINES

PACIFIC
OCEAN

South China Sea

MALAYSIA

Sulawesi
Sea

LIGITAN
SIPADAN

Molucca
Sea

Seram Sea

Banda Sea

IIIC

Arafura Sea

600 kilometres

© Robert Cribb 2009

TIMOR-
LESTE

IIIB

Timor Sea

Savu Sea

IIID

IIIA

Makassar St

Lombok St

II

MALAYSIA

MANGKAI

Natuna Sea

Karimata St

CHRISTMAS I.
(AUSTRALIA)

I
Sunda St

Malacca St

MALAYSIA

NICOBAR IS
(INDIA)

INDIAN
OCEAN

Indonesia's
archipelagic
base lines

Agreed north–south
archipelagic sea lanes

East–west routes proposed by Indonesia,
currently under normal regime

29

Indonesia is the world's largest archipelagic state and has the third largest EEZ in the world.[2]

Indonesia has, it must be noted, a number of disputes with its neighbours over the precise boundaries of some of its claims to maritime territory. In the most prominent of these, Indonesia and Malaysia both claim jurisdiction over a section of sea off the east coast of Kalimantan known as the Ambalat block; in 2005 tensions between them reached a peak when naval vessels from the two countries confronted one another in this area. Leaving aside disputes over what are, in relation to the whole, small patches of sea, however, the lines on a standard map of Indonesia are accepted by its neighbours. These lines, moreover, conform to the principles set out in the United Nations Convention on the Law of the Sea, ensuring that Indonesia's status as an archipelagic state has international recognition. But how did Indonesia become an archipelagic state in the first place? The answer to this question lies in events that began in the early 1950s, but before I examine those events I need to give a brief account of the claims to the sea made by the Netherlands Indies government, for it was an ordinance issued by the colonial government in 1939 that defined Indonesia's maritime territory in its first years as a nation-state.

## THE TERRITORIAL SEA AND MARITIME DISTRICTS ORDINANCE OF 1939

For most of the nineteenth century the government of the Netherlands Indies took little interest in defining the nature and extent of its jurisdiction over the seas adjacent to its land territory (Resink 1968: 120–21), but the arrival of Australian pearlers in the eastern part of the archipelago in the 1880s suddenly forced it to give this question a great deal of attention. In their search for pearl shell the Australians often came into conflict with villagers and rulers who had long laid claim to riches found on the reefs where the pearlers collected shells. The government immediately issued an ordinance making it illegal for foreigners to collect shell within the 'territorial waters of the Netherlands Indies' without the permission of the governor-general (Netherlands Indies 1893: No. 261). The ordinance did not define these waters but the government made it clear in other ways that it observed the three-mile limit championed by Great Britain, which, as the world's leading maritime power, had a vested interest in

---

2   The EEZs of the two states with larger EEZs, namely the United States and France, include waters around their overseas dependencies and territories (Churchill and Lowe 1999: 178).

restricting the maritime claims of other states to as narrow a band of sea as possible. Thus, each island had its own territorial waters extending out to three nautical miles from that island's low-tide line.

At the same time as it was dealing with angry Australian pearlers, the government conducted a survey of the various self-governing realms of the Indies to find out whether they asserted any sort of claim to the seas adjacent to their territories. Somewhat to its surprise, the government discovered that several rulers did indeed assert such claims. The sultans of Ternate and Tidore claimed the exclusive right to collect pearl shell, tripang and other highly valuable marine products in certain areas, while the *raja* along the east coast of Sumatra regarded the waters near their land territories as part of their 'domains'.[3] As a result of an application by a European mining company to dredge for tin in the waters next to Singkep island, the government also discovered that the sultan of Riau-Lingga claimed 'territorial rights' over the waters around his islands out as far as a person could see from the shore.[4] After lengthy and bitter debate both within the Indies and in The Hague, the government took the step in 1902 of denying the self-governing realms any possibility of having their own territorial waters (Teitler 1994: Ch. 2). From then on, the government declared, only the government in Batavia could exercise sovereignty over the sea. In fact, the government emphasized its authority in this respect by declaring in a statute concerning its agreements with these realms that 'the territory of [these] lands includes no sea territory of any kind' (Netherlands Indies 1919: No. 822). As 'a favour' (*gunst*) the government allowed some rulers to continue to exercise rights over the collection of marine resources, and in the 1920s it gave limited recognition to the claims of some villages, particularly in the eastern part of the archipelago, to adjacent waters, but jurisdiction over all waters within three miles of the coast of all the islands in the Netherlands Indies ultimately lay with the government.

During the early decades of the twentieth century the Netherlands Indies government, like other governments around the world, tried to expand the area of sea that fell within its jurisdiction by stretching the

---

3   Resumé van de Antwoorden der Betrokken Hoofden van Gewestelijk Bestuur op de Dezerzijdsche Circulaire van 4 April 1893 No. 1986 Betreffende het Visschen van Parelschelpen, Parelmoerschelpen en Tripang in de Territoriale Wateren in Nederlandsch-Indië [Summary of the Answers of the Relevant Heads of Regional Administration to the Circular dated 4 April 1893 No. 1986 concerning the Fishing of Pearl Shells, Mother-of-pearl Shells and Tripang in the Territorial Waters of the Netherlands Indies], openbaar verbaal, 18 September 1896, No. 48, Ministerie van Koloniën, Nationaal Archief, The Hague.

4   Resident of Riau to Governor-General, 30 May 1899, openbaar verbaal, 20 September 1902, No. 34, Ministerie van Koloniën, Nationaal Archief, The Hague.

meaning of the three-mile limit.[5] In 1905 it declared that a rock, reef or bank within six miles of the coast that fell dry at low tide could have its own territorial waters. During World War I, when the Netherlands was desperate to maintain its neutrality, both the home and colonial governments searched for a way to prevent foreign powers from constructing installations on rocks far out to sea. 'One asks', wrote the Minister of Foreign Affairs in 1915, 'whether a different policy should be followed in the case of an archipelago that geographically and economically forms a more or less cohesive whole' (J. Loudon, quoted in Teitler 1994: 77). Eventually, in 1927, the government issued an ordinance that declared that the 'sea territory of the Netherlands Indies' included the waters within three nautical miles of rocks, reefs and banks 'in the Netherlands Indies archipelago' (Netherlands Indies 1927: No. 144). Suddenly, at a multitude of points scattered throughout the archipelago, specks of sea, each covering about 28 square nautical miles, came under the government's jurisdiction. In the Territorial Sea and Maritime Districts Ordinance of 1935, however, the government retreated from this extension of its maritime territory because of doubts about its standing in international law. In fact, as far as rocks, reefs and banks were concerned, this ordinance claimed even less than the 1905 ordinance had, for it declared that such a feature had to be within three nautical miles of an island in order to generate territorial waters (Netherlands Indies 1935: No. 497).[6]

The 1935 ordinance did give the government greater authority *within* its maritime territory. Most notably, in an attempt to curtail surveillance activities by Japanese naval and fishing vessels, it gave the governor-general the power to designate certain areas as 'maritime districts' (*maritieme kringen*) within which far more restrictions applied than elsewhere. For example, the ordinance prohibited anyone who was not a native of the Indies from fishing in maritime districts and banned hydrographic surveys and photography in these waters without the government's permission. But of course it was only within its territorial waters (and the

---

5   This and the following paragraphs provide but a sketch of an immensely complicated history. See Teitler (1994) for a thorough account.

6   The ordinance used the term 'territorial sea' for the band of waters around each island and extending out from lines across the entrances to bays and estuaries, and the term 'territorial waters' to encompass both the territorial sea *and* internal waters, namely waters on the landward side of lines drawn across the entrances to bays and estuaries and on the landward side of lines drawn around clusters of islands where those islands were within six nautical miles of one another. The term 'territorial sea' came to be used in the various UN law of the sea conventions in the same sense as it was used in the 1935 ordinance, but throughout this paper I use the term 'territorial waters' as the equivalent of 'territorial sea' in that ordinance and later conventions.

internal waters of bays and estuaries) that the government could apply these restrictions. The maritime space directly under the jurisdiction of the Netherlands Indies government in 1935 was probably no greater than it had been at the beginning of the century.

The Territorial Sea and Maritime Districts Ordinance of 1939 (Netherlands Indies 1939: No. 442), issued on the eve of the outbreak of war in Europe, turned out to be the colonial government's final attempt to define its maritime territory. This ordinance amended the 1935 ordinance in many ways but it left in place the fundamental principle that, except in a few special cases, the territorial waters of the Indies were to be measured three nautical miles out from the low-tide line of each individual island. This meant that nearly all of the sea between the islands making up the Netherlands Indies had the status of international waters or high seas. Passenger, cargo and fishing vessels and even warships from anywhere in the world were as free to sail through these waters as if they were traversing one of the great oceans. Equally, to look at the same scene from the point of view of someone living in the Netherlands Indies, a vessel sailing between, for example, Java and Borneo, Sulawesi and the Aru Islands or even Java and Bawean travelled through international waters for much of its journey. By the 1930s more and more states around the world were becoming dissatisfied with the three-mile limit, but for the most part this dissatisfaction simply took the form of claiming, or threatening to claim, somewhat broader territorial waters.

## THE JUANDA DECLARATION

Independent Indonesia was not born as an archipelagic state. During the debates of the Committee for the Investigation of Independence near the end of the Japanese occupation, Muhammad Yamin argued that the seas between the terrestrial components of Indonesia (in which he hoped to include the Malay Peninsula) were as much a part of Indonesia as the land and that the principle of the freedom of the seas threatened Indonesia's sovereignty and security, but he did not go on to propose that the government of an independent Indonesia lay some form of claim to all the waters of the archipelago. Instead, he suggested closing narrow straits to foreign ships unless they had been granted a permit or flew the flag of a country with which Indonesia had a treaty allowing that country's ships to pass through. He also appears to have suggested a slight widening of the country's territorial waters (Yamin 1959–60: 134). That, it appears, was as far as the discussion went at this time.

Neither the 1945 nor the 1950 constitution made any reference to Indonesia having any form of jurisdiction over the waters between the islands. But there is evidence in the archives of the Australian Department of External Affairs that the newly independent government had already begun to look at the sea in a new way in the early 1950s. Australian officials were aware as early as 1953 that the Indonesian government was considering ways of extending its jurisdiction over a much greater expanse of sea. One source states that the government was considering the possibility of 'defining territorial waters as the waters between the islands with a sixty mile limit',[7] while another notes that 'Indonesia was contemplating the adoption of a coast to coast line which would in effect make' the seas of Indonesia 'closed waters for all purposes'.[8] These sources provide only one hint of what may have prompted these ideas, and that is that Indonesian officials were concerned about the effect the return of Japanese fishing vessels to the archipelago might have on stocks of fish and pearl oysters. In Australia itself, fear of the devastation Japanese pearlers might have on the stocks of oysters on reefs far from the Australian coastline (and in turn the impact that would have on the Australian pearling industry) led the government to declare sovereignty over the Australian continental shelf in 1953. Australian documents reveal that the Indonesian government had similar fears and considered issuing a similar declaration. Indonesian sources make it clear, however, that by 1956 Japanese fishing was just one of several problems that were prompting the government to consider ways of extending its maritime jurisdiction. Foremost among these was the question of how to wrest western New Guinea from Dutch control. It was at this point that events took place very rapidly.

The starting point for these events was an interdepartmental committee set up by Prime Minister Ali Sastroamijoyo in October 1956 to revise or replace the Territorial Sea and Maritime Districts Ordinance of 1939.[9] A memorandum from the minister of defence had prompted the prime minister to set up the committee, but several other ministries, including justice, foreign affairs, transport and agriculture (which included fisheries), all had an interest in the issue. Due in large part to the many changes in its membership, this committee — and its two subcommittees dealing with legal and technical issues — moved at a slow pace, taking about 14 months to prepare the draft of a new law. At least from the sketchy record available in published sources, it is clear that the committee considered a

---

7  Third Secretary (R.R. Fernandez) to Minister, 18 September 1953, Department of External Affairs, National Archives of Australia, A11604, 606/3/2.

8  For Assistant Secretary, March 1954, Department of External Affairs, National Archives of Australia, A1838, 3034/10/4/1.

9  In the following account I owe a general debt to Leifer (1978), Djalal (1996), Lapian (2007) and Danusaputro (1980), the first three of which led me to other sources that I have used as well.

number of possible ways of extending the government's control over the waters of the archipelago. It is important to keep in mind that by the time it began its deliberations several states had unilaterally extended their jurisdiction over the waters and seabeds adjacent to their land territories. Most famously, in 1945 President Truman issued a proclamation giving the United States exclusive jurisdiction over resources on and under the seabed of the continental shelf adjacent to its coasts, a step that other states, including Australia, soon followed. As far as what happened in Indonesia is concerned, however, two other precedents were far more significant. It is important to say a few words about each of these if we are to judge the significance of events that took place in Indonesia.

The first of these precedents was the Anglo-Norwegian Fisheries case of 1951 in which the International Court of Justice (ICJ) ruled on the legality of the straight baselines joining the outermost points of the thousands of islands along the coast of Norway that the Norwegian government had used to delimit its territorial waters. The feature of these straight baselines that prompted the British government to challenge Norway in the court was the fact that they gave Norway jurisdiction over far more maritime territory than if the Norwegians had followed the practice that the British had long advocated of measuring territorial waters three miles out from the low-tide line of the mainland and each and every island. Pointing out that neither Britain nor any other state had objected between 1869 (when the Norwegian government first applied the baselines) and the 1930s, and that the Norwegians had a close historical and economic connection with the waters concerned, the court ruled that the baselines were not contrary to international law (Churchill and Lowe 1999: 33–5).

Around the world officials and lawyers instantly understood that this ruling might have much broader implications. Asked by the US Secretary of State to give his view on the ICJ's decision, on 22 June 1952 the Secretary of the Navy declared that 'any action which tends to restrict free navigation of the high seas by recognizing sovereignty over territorial waters in excess of three miles is contrary to United States security interest' (US Department of State 1983: 1,660). He referred specifically to Indonesia as one of the places where the ICJ's ruling might have unfortunate consequences for the United States, though his letter shows that he had little idea of how far the Indonesians might take the straight baseline principle.

The other precedent that the committee drafting a new law considered was a *note verbale* issued by the government of the Philippines in March 1955:

All waters around, between and connecting different islands belonging to the Philippines Archipelago, irrespective of their width or dimension, are necessary appurtenances of its land territory, forming an integral part of the national or inland waters, subject to the exclusive sovereignty of the Philippines (Evensen 1958: 299).

It is notable that this proclamation left open the question of exactly how these waters would be delimited. It is possible that the Philippine government took the Anglo-Norwegian Fisheries case as its starting point and considered applying straight baselines to the archipelago, but there is no sign of this in the proclamation itself. Nevertheless, it was clear to everyone at the time that the government had in these few words vastly extended the territory over which the Philippines claimed sovereignty.

Despite these precedents, the Indonesian interdepartmental committee rejected, at some stage in its proceedings, the possibility of claiming either sovereignty or a more limited form of jurisdiction over the waters between the islands making up Indonesia. It appears to have specifically rejected the possibility of applying the principles of the Anglo-Norwegian Fisheries case to the archipelago as a whole. Instead, the most important article in the draft law that it handed to cabinet on 7 December 1957 extended Indonesia's territorial waters, still measured from the low-tide line of each island, from three to 12 nautical miles. Although in another article the government reserved the right to extend the country's maritime jurisdiction still further at some later date, the draft law was, as Munadjat Danusaputro (1980: 101) remarks, largely an updated version of the 1939 ordinance. Seen from our vantage point more than half a century later, the draft law is indeed a remarkably timid document, but the committee had what it regarded as a compelling reason not to take a bolder course. Its main concern was that at that stage the government simply did not have the military and other resources needed to enforce its jurisdiction over the vast area that would suddenly become Indonesian territory if it were to draw straight baselines around the archipelago. In light of the weakness of the Indonesian navy at the time, this concern made perfect sense.

By the time the committee finally completed its deliberations, however, the political situation in Indonesia had changed greatly from when it began its work in 1956, making the government receptive to much bolder proposals. Politicians in regions outside Java were becoming increasingly dissatisfied with policies that they regarded as favouring Java over the rest of the country, army commanders in a number of regions had declared martial law, President Sukarno had in turn declared martial law over the whole country and set up a 'business cabinet' headed by Juanda Kartawijaya, labour unions had taken over Dutch enterprises, and tensions between Indonesia and the Netherlands over western New Guinea had increased greatly. These tensions were heightened even further when, in what Indonesian accounts regard as a blatant display of force,

Dutch warships sailed through the Java Sea and seas in the eastern part of Indonesia on their way to western New Guinea. What infuriated some Indonesian politicians was the fact that under the 1939 ordinance the Indonesian government could do nothing to prevent the warships from passing through these waters. Sometime in November 1957, Chairul Saleh, the Minister of Veteran Affairs and a close confidant of Sukarno, asked a young member of the interdepartmental committee to whom he was related by marriage why the committee was going about its work so slowly and timidly. That committee member was Mochtar Kusumaat-maja, who had been appointed to the committee only in August after returning to Indonesia from studies at Yale Law School. Mochtar was, by his own account, somewhat taken aback by Chairul's displeasure with the committee and his suggestion that Indonesia would never have gained its independence had the leaders of the nationalist movement been so fainthearted. Chairul challenged Mochtar to find some legal instrument that would allow the Indonesian government to prevent Dutch warships from passing through the Java Sea. After Chairul helped him get leave from the department where he worked, Mochtar headed off to the relative tranquillity of Bandung to come up with a way of dealing with this problem (Kusumaatmadja 1993).

Mochtar's solution was to draw straight baselines between the outermost points of the outermost islands on a map of Indonesia in a primary school atlas that he had brought with him. 'It was precisely at this moment', writes Danusaputro (1980: 105), 'that Indonesia tried to apply the "archipelagic principle" to itself for the first time'.[10] The waters within these baselines would become Indonesia's internal waters, over which the government would exercise full sovereignty, while the country's territorial waters would be measured out from the baselines. Mochtar appears to have devoted considerable thought to how this step could be justified under international law. Danusaputro's *Tata Lautan Nusantara* contains a document dated 9 December 1957 that appears to have been written by Mochtar (Danusaputro 1980: 140–46). Whoever wrote it, it was this document that provided the initial justification for the archipelagic principle.

---

10  According to Djalal (1996: 30), Mochtar 'stumbled upon the Anglo-Norwegian Fisheries case' at this time. My own view is that this is unlikely, as it seems clear from Danusaputro's (1980: 100-1) account that the interdepartmental committee (of which he was a part) was fully aware of it, as, it would appear, were most politicians and lawyers around the world who were dealing with the question of territorial waters. What he was not aware of, as Leifer (1978: 20) notes, was Evensen's paper, dated 29 November 1957, which was to provide much of the rationale for the archipelagic principle in international law (see Evensen 1958).

After methodically considering and then rejecting various alternatives, such as extending the width of territorial waters and encircling the territorial waters of each island with a contiguous zone within which the government could exercise jurisdiction over specific matters such as customs and quarantine, the document argued that the application of baselines provided the only means by which the government could meet its objectives. Moreover, the document insisted, this step would be entirely consistent with international law. While noting that some of the baselines would be longer than the longest of those in the Norwegian system of baselines, the document argued that the basic principle of the ICJ's decision in the Anglo-Norwegian Fisheries case applied just as well to Indonesia as it did to Norway: the waters of the Indonesian archipelago were historically, economically and geographically an integral part of Indonesia. Indonesia was, the document declared, 'a single united archipelago'. As part of its considerations, the document also dealt with the issue of the width of the territorial waters that would form a band around the baselines. Interestingly, the document made the point that once Indonesia had laid down baselines encircling the whole country, the width of these waters did not particularly matter. There was, it argued, no need to claim more than the traditional three nautical miles; what mattered was the baselines, for it was the baselines that would define the archipelago.

When the cabinet met on the evening of 13 December 1957, it began by focusing on the question of how to deal with the Netherlands, which had apparently stepped up its naval presence in the Java Sea following the nationalization of Dutch companies on 3 December.[11] In this context cabinet turned its attention to the interdepartmental committee's report advocating an extension of Indonesia's territorial waters to 12 nautical miles and Mochtar's proposal to draw baselines around the whole archipelago. At first, according to Danusaputro's account, the debate concentrated entirely on what approach would best enable the government to deal with the particular problems it faced, not only the challenge posed by the Dutch but also the threat of regional rebellion and the need to protect Indonesia's fisheries and other marine resources, but at some stage Juanda suggested that the cabinet focus instead on the broader question of what approach would serve Indonesia best in the long term. It was, it appears, at this point that the cabinet embraced Mochtar's proposal. Its great attraction from the point of view of cabinet members was that it made Indonesia whole. Indonesian politicians of all parties appear to have become increasingly irritated during the 1950s by the notion that Indonesia was full of 'holes' made up of international waters; these holes not only made it extremely difficult for the navy and other agencies to

---

11   Danusaputro (1980: 105-7, 1993) gives accounts of this meeting.

enforce Indonesian law but also deeply offended Indonesians' sense of nationhood. Applying baselines to the archipelago removed these holes and gave the government the legal instrument it needed to deal with the multitude of problems it faced in the seas of the archipelago. Most important, it created at least the image of a united Indonesia. Referring in part to US involvement in the regional unrest of this time, Mochtar described some years later how important this image had been in the cabinet's thinking:

> [As] the politicians saw the country falling apart, they said, 'We must have a concept that shows these simple people physically that we are one'. ... The people had to be shown in simple symbols that Indonesia was one. We had just gotten our independence, and we had all these big boys interfering, trying to keep us apart because they had their own designs. So this archipelago principle seemed to be a good thing for the important political unity of Indonesia (Kusumaatmadja 1973: 176).

Having drawn straight baselines around the archipelago, cabinet might well have stopped at this point, but it took the further step of expanding the width of Indonesia's territorial waters to 12 nautical miles, a width that had recently been adopted by several other states. In effect, cabinet accepted the interdepartmental committee's proposal to extend Indonesia's territorial waters from three to 12 nautical miles but only after adopting the fundamentally different approach of measuring these waters out from baselines encircling the whole archipelago rather than from the low water line of each island. After discussing the reaction it was likely to receive from the Netherlands and the major maritime powers, cabinet decided to issue a statement titled 'Government Declaration concerning the Water Areas of the Republic of Indonesia', soon generally known as the Juanda Declaration.

The Juanda Declaration of 13 December 1957 is a model of simplicity. After noting the uniqueness of the Indonesian archipelago and the need to regard the islands of Indonesia and the seas between them 'as one total unit', and asserting that the article in the 1939 ordinance defining the extent of the country's territorial waters 'is no longer in accordance' with these considerations because 'it divides the land territory of Indonesia into separated sections', it declares that:

> ... all waters surrounding, between and connecting the islands constituting the Indonesian state ... are integral parts of the territory of the Indonesian state and, therefore, parts of the internal or national waters which are under the exclusive sovereignty of the Indonesian state.

It attempts to reassure the maritime powers (and protect Indonesia's overseas trade) by stating that 'innocent passage of foreign ships in these internal waters is guaranteed as long as it does not violate or interfere with the sovereignty and security of Indonesia'. It then neatly defines both Indonesia's territorial waters and, implicitly, its internal waters: 'The delimitation of the territorial sea (the breadth of which is 12 miles) is measured from baselines connecting the outermost points of the islands of Indonesia'. It concludes by noting that the government planned to enact these provisions 'as soon as possible' and that its 'position will be maintained' at the United Nations Conference on the Law of the Sea scheduled to begin in Geneva in February 1958.[12]

As I have suggested, the notion of applying baselines to enclose an archipelago had precedents. Indeed, the author of the 9 December document had been at pains to emphasize these precedents when he advocated applying this notion to Indonesia. But the Juanda Declaration went much further. It applied baselines not to a group of islands that formed the fringe of a state, as the Norwegian government had done, nor to a small group of islands, as the Cuban government had to the Canarreos Archipelago, but to a massive archipelago. Moreover, the declaration defined far more clearly than the Philippines' *note verbale* the principle by which the government would delimit the waters under its sovereignty. It marked, as Indonesian accounts assert, a radical development in the law of the sea. Most importantly, to look at the declaration within the context of Indonesia's history, it laid the basis of the archipelagic state.

## THE 25-YEAR 'STRUGGLE'

The reaction of the Netherlands and the major maritime powers to the Juanda Declaration was swift and strong. In their view the declaration violated international law and constituted a major encroachment on freedom of movement on the world's oceans. The statement in the declaration granting foreign vessels innocent passage failed to assuage them. Innocent passage did not give vessels the same degree of freedom as that enjoyed in international waters. More fundamentally, the very use of the term implied the government's sovereignty over the waters in question. By the middle of January 1958, the United States, the United Kingdom, Australia, New Zealand, France and Japan as well as the Netherlands had handed the Indonesian government protest notes expressing their views on the declaration. Somewhat patronizingly, some of the notes suggested that, because the announcement had taken the form of a declaration rather than actual legislation, perhaps the Indonesian government might decide not to pursue the matter any further. All governments

---

12  See Syatauw (1961: 173-4) for an English version of the Juanda Declaration and Danusaputro (1980: 135-6) for the Indonesian text.

made the point in their notes that the declaration would not be regarded as binding as far as their own ships were concerned. In the West newspapers joined in the condemnation; *The Times*, for example, referred to the declaration as 'the new piracy' (18 December 1957: 9).

At the Law of the Sea Conference, the head of the US delegation attacked the declaration so vehemently that his Indonesian counterpart, Subarjo Joyoadisuryo, took the first possible opportunity to defend his government's position. In response to the US position that the declaration would 'subtract from the high seas, the common property of all nations', Subarjo argued that 'the fact that the seas were the common property of all nations did not preclude the possibility of a special regime for archipelagos of a unique nature'.[13] As it turned out, the conference devoted hardly any time to discussing such a possibility. In the months following the declaration a small number of states, most notably the Soviet Union, expressed their support for the Indonesian position but for the most part it met with nothing but condemnation.

Faced with such opposition, the government appears to have decided not to rush into enacting the declaration, but with the approach of the second Law of the Sea Conference in 1960 it took the pre-emptive step of preparing legislation. The result was Law No. 4/1960 on Indonesian Waters, dated 18 February 1960. Included in the law were the coordinates of all the points used to define the ends of each of the 196 baselines encircling the country and a map showing the baselines and the band of territorial waters extending 12 nautical miles out from the baselines. It is important to note that, despite the clarity of the Juanda Declaration, the governments of other countries had not known exactly how the baselines would be drawn. They were fully aware that the longer the baselines, the greater the area that would be included within Indonesia's internal waters. As it turned out, the two longest baselines, one stretching across the northern entrance to the Molucca Sea and the other reaching from the northern coast of western New Guinea to Mapia Island, were each about 123 nautical miles long, far longer than any in the Norwegian system. In any case, with these baselines the government laid the basis for Indonesia's maritime boundaries. Indeed, leaving aside some minor changes, this map provides the lines we see on present-day maps.[14] Significantly,

---

13  See United Nations (1958: 25, 44). These are summary rather than verbatim records of the proceedings.

14  For a detailed analysis of Indonesia's baselines, see Forbes (1995). For a copy of the 1960 map, see Syatauw (1961: after 168). The most important change in recent years concerns the waters between Sumatra and Kalimantan. In the 1960 map an area to the north of Belitung remained as international waters; recent revisions to the baselines have incorporated that area into Indonesia's archipelagic waters.

the map included western New Guinea even though that territory was still under Dutch control.

Like the Juanda Declaration, Law No. 4/1960 was subjected to a round of condemnation from the Netherlands and the major maritime powers, and the Indonesian government had no more success in persuading the second Law of the Sea Conference to address the question of archipelagic baselines than it had had at the first conference. Undeterred, in 1962 the government issued a regulation (No. 8/1962) spelling out in some detail the conditions under which foreign ships might sail through Indonesian waters. Alarmingly for foreign governments, the government declared that unlike in territorial waters, where innocent passage was a right, 'innocent passage in the inland sea is a facility purposely granted by Indonesia' and that because of this the government reserved the right to withdraw those 'facilities'. Even more alarming was a requirement that foreign warships give notice of their intention to pass through Indonesian waters (Leifer 1978: 24, 187–8). The days when ships had enjoyed the freedom of the sea in this part of the world were, it appeared, over.

These dramatic steps taken by the Indonesians presented the major maritime powers with a profound dilemma. On the one hand, they objected, as a matter of deeply held principle, to the steps taken by Indonesia. They particularly balked at the requirement that they notify the Indonesian government whenever one of their warships was about to pass through an area claimed by Indonesia, for any such notification would amount to an acknowledgment of the government's claims. On the other hand, they dreaded the prospect of inflaming the political situation in Indonesia and giving the increasingly powerful Indonesian Communist Party (Partai Komunis Indonesia, or PKI) more grounds for attacking the West. The archives are full of files in which officials desperately debate how to deal with Indonesia. This was, it is worth adding, not a new problem. In 1954 the Australian government had called off a proposed naval exercise in the Arafura and Banda seas for fear of upsetting the Indonesians; implicit even at that time was an acknowledgment that the Indonesians had *some* kind of right over the seas near their islands.[15] But by the early 1960s this fear – and this implicit acknowledgment – had become much stronger. As Toh Boon Kwan (2005) shows in the case of British naval movements through the archipelago during the Confrontation period, the Western powers might express the strongest objections to what they saw as infringements on their freedom of movement, but in the end they always acquiesced, if only partially, to Indonesian demands.

---

15   Correspondence in National Archives of Australia, A1838, 3034/10/4/1.

Indonesians often refer to their long campaign to gain international recognition of the archipelagic principle as a 'struggle' (*perjuangan*),[16] the word commonly used to describe the fight for independence after World War II. Whether they knew it or not, their greatest ally in this struggle in the first few years after the Juanda Declaration was the fear Western governments had of doing something that might tip Indonesia into chaos or, worse, turn it communist.[17] Even after the events of 1965–66, Western governments held back from directly challenging Indonesia on maritime issues because of their desire to build good relations with the New Order regime. As a result, with each passing year, with each notification, however informal, of the arrival of a Western warship, the archipelagic principle was gradually becoming a de facto part of the law of the sea. As O'Connell explains in an article on archipelagos, in customary international law 'the advantage lies with the party which acts and the disadvantage with the party which must demonstrate that the action is illegal' (O'Connell 1971: 63). The Indonesian government made full use of that advantage.

Beginning in the late 1960s, the Indonesian government took a series of steps aimed at gaining formal recognition of Indonesia as an archipelagic state. It began by negotiating a series of maritime boundary agreements with neighbouring states, of which those with Singapore, Malaysia and Australia were the most important. It appears that in the negotiations leading up to these agreements the government was prepared to compromise when needed in order to gain implicit recognition of the baselines as the starting point for whatever claim Indonesia was making. It is notable, for example, that as a result of an agreement with the Singapore government, a small section of Singapore's territorial sea is located on the Indonesian side of Indonesia's baselines (Forbes 1995: 24–5). In this way, it appears, the government slowly gained the confidence and support of neighbouring countries.

In the meantime, Indonesian diplomats began putting forward their government's position at academic conferences dealing with the law of the sea, making the case that the Indonesian government was simply taking steps to ensure the country's security and territorial integrity just as any other government would. They used meetings of such bodies as the Asian African Legal Consultative Committee to gain support from newly independent states. Finally, in 1969, they began taking part in planning for a third United Nations Conference on the Law of the Sea (LOSC III).

---

16  See, most notably, Djalal (1979).
17  Another ally was division among the Western countries about how to deal with Indonesia. As in other matters, the US government failed to resist as strongly as the Netherlands government had hoped it would.

By the late 1960s, as Hasjim Djalal (2001: 344) explains, the great increase in the number of African and Asian states as a result of decolonization, the growing awareness of the threat to the marine environment posed by supertankers, developments in the technology for extracting oil from under the sea, advances in methods of finding and catching fish and Cold War tensions had all contributed to a widespread view that the time had come to define much more clearly the nature of the jurisdiction states exercised over adjacent waters. When the conference finally began its work in 1973, Indonesian diplomats took the lead in pushing the interests of archipelagic states.

It is important to note that throughout these years (and well beyond) Indonesia was represented by the same two diplomats, namely Mochtar Kusumaatmaja, whose work in November–December 1957 had, as we have seen, laid the basis for the Juanda Declaration, and Hasjim Djalal, who had begun working with Mochtar in the early 1960s soon after completing a doctoral dissertation on the limits of territorial waters. Files in the Australian archives reveal their ability to gain the trust of those they were negotiating with, while a casual perusal of the proceedings of academic conferences on the law of the sea indicates that they rarely missed a chance to present the Indonesian position to a wider audience. Having built up a worldwide network of official and academic associates, Mochtar and Djalal were in a strong position to put forward the view of Indonesia and the other archipelagic states at LOSC III.

The meetings of LOSC III extended over nine gruelling years. With tributes to Simón Bolívar at the opening session in Caracas and laments over the death of Chairman Mao at a session in 1976, the atmosphere of LOSC III was very different from that of the earlier conferences. Even so, the Indonesian delegation still faced immense challenges. In the early sessions, and also in the scholarly literature of this time, there was extensive debate about the concept of an archipelagic state. How many islands did it take to form such a state? How close did these islands have to be to one another? Should a country made up of a few tiny, widely scattered islands qualify as an archipelagic state? At the same time, there was debate about the nature of the jurisdiction an archipelagic state might have over the waters between its islands. Were these waters internal waters where the state's sovereignty was absolute? Were they territorial waters where the state was bound to allow foreign ships the right of innocent passage? Were they much like a contiguous zone where the state's jurisdiction was limited to specific functions such as the regulation of the exploitation of resources or the enforcement of customs and quarantine laws? Or should some special regime be created for these waters? Should these waters in fact be called 'archipelagic waters'? If so, should that term become part of international law?

Fundamental though these questions were, the most striking aspect of the negotiations concerning archipelagos at LOSC III was that for the most part they led to a confirmation of legislation and regulations that the Indonesian government had had in place since the early 1960s. By the middle of 1977 the governments of the United States and the other maritime powers had indicated that they would be prepared to accept the archipelagic concept in principle. As it turned out, the US government tried to delay the proceedings at LOSC III soon after the Reagan administration came into office in 1981, but by then the great majority of states involved in the negotiations had committed themselves to reaching an agreement.

The culmination of the negotiations that took place at LOSC III was the United Nations Convention on the Law of the Sea, which was opened for signing in December 1982 and came into force in 1994 after being ratified by the required number of states. Of most importance to the Indonesian government was Part IV of the convention, 'Archipelagic States'. Part IV defined the concept of an archipelagic state, set certain limits on the length of baselines, declared that archipelagic states had sovereignty over the 'archipelagic waters' enclosed by the baselines, made it clear that archipelagic states were expected to take into consideration the interests of any neighbouring states affected by the drawing of baselines, and set out a procedure for the establishment of 'archipelagic sea lanes' through which foreign vessels would be able to pass in 'normal mode'.

The influence of Indonesian diplomats can be seen in virtually every aspect of Part IV. To take just one example, the article dealing with the length of baselines contained the seemingly arcane stipulation that the 'length of such baselines shall not exceed 100 nautical miles, except that up to 3 per cent of the total number of baselines enclosing any archipelago may exceed that length, up to a maximum length of 125 nautical miles' (article 47(2)). The article was written in this way precisely to give recognition to the baselines that Indonesia had already proclaimed in 1960. It must be emphasized that the Indonesian delegates did have to make some compromises. Most notably, they agreed to the regime of archipelagic sea lane passage even though it permitted, among other things, submarines to pass through an archipelago while submerged as long as they stayed within a designated sea lane. Nevertheless, the Convention on the Law of the Sea was a triumph of Indonesian diplomacy. The icing on the cake came in 1988 when the US government, which had not signed the convention, signed a double taxation agreement with Indonesia that was accompanied by an exchange of notes in which the United States recognized Indonesia as an archipelagic state, though with the pointed addition of words declaring that it did so on the understanding 'that Indonesia respects international rights and obligations pertaining to transit of the Indonesian archipelagic waters' (Roach and Smith 1996: 212).

One further point about the convention is worth making. In 1980 the Indonesian government joined the worldwide rush to declare EEZs. Because it used its archipelagic baselines as the starting point from which to measure its EEZ, an immense area of ocean came under Indonesian jurisdiction as a result of this move. Again, the Convention on the Law of the Sea gave recognition to a step Indonesia had already taken, but in this case Indonesia was just one of a multitude of states laying claim to more of the sea. In fact, the Indonesian government was one of the last in Southeast Asia to claim an EEZ. During the early sessions of LOSC III, Mochtar explained to an Australian official, 'the most important thing for Indonesia was to win the struggle for international recognition of the archipelagic concept'. It was only in the middle of 1977, when neighbouring states had made it clear that they would soon announce EEZs and when, as noted, the government was confident that its position on archipelagic states would prevail at LOSC III, that Indonesian officials began to turn their attention to the possibility of declaring an EEZ.[18]

The Indonesian government still faces a number of contentious issues concerning its maritime space. The most important of these concerned Indonesia's maritime borders with Malaysia and the reluctance of the government to set up an east–west archipelagic sea lane in addition to the three north–south sea lanes it proclaimed in the 1990s.[19] Nevertheless, however these matters are resolved, the fundamental principles embodied in the Juanda Declaration have been firmly in place since 1982. Indonesia is universally recognized as an archipelagic state.

## LOOKING BACK

The Juanda Declaration is rarely mentioned in general histories of Indonesia written in English. Perhaps this is not surprising, for it had little immediate effect on the tumultuous course of events during 1957–58. But the Juanda Declaration does deserve our attention. Along with the basic principle that Indonesia would be made up of the islands once governed by the Netherlands, it was this declaration that defined Indonesia's territory. It is therefore understandable that for many Indonesians the Juanda Declaration looms large in the nation's history. Indeed, it came to be por-

---

18   Australian Embassy, Jakarta, to Canberra, 27 July 1977 (reporting conversation with Mochtar), and 28 September 1977 (reporting conversation with Hasjim Djalal), National Archives of Australia, A1838, 1734/83, Part 1.

19   On the question of Indonesia's archipelagic sea lanes, see Bateman (2006).

trayed as the natural, predestined realization of Indonesia's nationhood. From a historiographical perspective such a view of the Juanda Declaration is highly questionable. It downplays the creativity and boldness of those most involved in imagining it and making it a reality, and it gives the declaration a purpose that it almost certainly did not have at the time. Nevertheless, with the benefit of hindsight we can acknowledge that it was a momentous event. From that point on, first in Indonesian law but later in international law as well, the territorial components of Indonesia formed one physical whole.

## ACKNOWLEDGMENTS

I am grateful to Robert Cribb, Bob Elson, Viv Forbes and Toh Boon Kwan for their very helpful comments on earlier versions of this chapter.

## REFERENCES

Bateman, Sam (2006), 'Security and the law of the sea in East Asia: navigational regimes and exclusive economic zones', in D. Freestone, R. Barnes and D.M. Ong (eds), *The Law of the Sea: Progress and Prospects*, Oxford University Press, Oxford and New York, Chapter 18.

Churchill, R.R. and A.V. Lowe (1999), *The Law of the Sea*, third edition, University of Manchester Press, Manchester.

Danusaputro, St. Munajat (1980), *Tata Lautan Nusantara dalam Hukum dan Sejarahnya* [The Nusantara Sea Regime in Law and History], Binacipta, Bandung.

Danusaputro, St. Munajat (1993), 'Negara kepulauan Indonesia gagasan Chairul Saleh' [The archipelagic state of Indonesia was Chairul Saleh's concept], in Dra. Irna H.N. Soewito (ed.), *Chairul Saleh: Tokoh Kontroversial* [Chairul Saleh: A Controversial Figure], Tim Penulis, Jakarta, pp. 229–36.

Djalal, Dino Patti (1996), *The Geopolitics of Indonesia's Maritime Territorial Policy*, Centre for Strategic and International Studies, Jakarta.

Djalal, Hasjim (1979), *Perjuangan Indonesia di Bidang Hukum Laut* [Indonesia's Struggle in the Law of the Sea], Penerbit Binacipta, Jakarta.

Djalal, Hasjim (2001), 'Deklarasi Djuanda menyatukan kita' [The Juanda Declaration united us], in Awaloedin Djamin (ed.), *Pelawan Nasional: Ir. H. Djuanda, Negarawan, Administrator, dan Teknokrat Utama* [A National Hero: Ir. H. Juanda, Eminent Statesman, Administrator and Technocrat], Penerbit Kompas, Jakarta.

Evensen, Jens (1958), 'Certain legal aspects concerning the delimitation of the territorial waters of archipelagos', in *Official Records: United Nations Conference on the Law of the Sea*, Volume 1, Geneva.

Forbes, Vivian Louis (1995), *Indonesia's Maritime Boundaries*, Malaysian Institute of Maritime Affairs, Kuala Lumpur.

Kusumaatmadja, Mochtar (1973), 'Supplementary remarks', in L.M. Alexander (ed.), *The Law of the Sea: Needs and Interests of Developing Countries: Proceedings of the Seventh Annual Conference of the Law of the Sea Institute, University of Rhode Island, Kingston, Rhode Island, June 2–29, 1972*, University of Rhode Island, Kingston, pp. 172–7.

Kusumaatmadja, Mochtar (1993), 'Sekelumit pengalaman bersama Bung Chairul Saleh' [Some experiences with Bung Chairul Saleh], in Dra. Irna H.N. Soewito (ed.), *Chairul Saleh: Tokoh Kontroversial* [Chairul Saleh: A Controversial Figure], Tim Penulis, Jakarta, pp. 173–81.

Lapian, Adrian B. (2007), 'Lima puluh tahun wilayah Republik Indonesia' [Fifty years of the territory of the Republic of Indonesia], paper presented to Dinamika Kemaritiman dalam Perspektif Sastra dan Sejarah, Seminar Internasional [International Seminar on the Dynamics of Maritime Affairs from the Perspective of Literature and History], Fakultas Sastra, Universitas Diponegoro, Semarang, 15 December.

Leifer, Michael (1978), *Malacca, Singapore, and Indonesia*, Sijthoff & Noordhoff, Alphen aan den Rijn.

Netherlands Indies (1893, 1919, 1927, 1935, 1939), *Staatsblad van Nederlandsch Indië* [Official Gazette of the Netherlands Indies], The Hague.

O'Connell, D.P. (1971), 'Mid-ocean archipelagos in international law', *British Yearbook of International Law*, 45: 1–77.

Resink, G.J. (1968), *Indonesia's History between the Myths*, W. Van Hoeve, The Hague.

Roach, J. Ashley and Robert W. Smith (1996), *United States Responses to Excessive Maritime Claims*, second edition, M. Nijhoff, The Hague and Boston.

Syatauw, J.J.G. (1961), *Some Newly Established Asian States and the Development of International Law*, Martinus Nijhoff, The Hague.

Teitler, G. (1994), *Ambivalentie en Aarzeling: Het Beleid van Nederland en Nederlands-Indië ten aanzien van Hun Kustwateren, 1870–1962* [Ambivalence and Hesitation: The Policy of the Netherlands and Netherlands Indies with Respect to Their Coastal Waters, 1870–1962], Van Gorcum, Assen.

Toh Boon Kwan (2005), 'Brinkmanship and deterrence success during the Anglo-Indonesian Sunda Straits crisis, 1964–1966', *Journal of Southeast Asian Studies*, 36: 399–417.

United Nations (1958), *Official Records: United Nations Conference on the Law of the Sea*, Volume 3 (First Committee), Geneva.

US Department of State (1983), *Foreign Relations of the United States, 1952–1954; General: Economic and Political Matters*, Volume I, Part 2, US Government Printing Office, Washington DC, available at http://digicoll.library.wisc.edu/FRUS/.

Yamin, Muhammad (1959–60), *Naskah-persiapan Undang-undang Dasar 1945* [Preparatory Documents on the 1945 Constitution], Volume 1, Jajasan Prapantja, Jakarta.

# 3    INDONESIA'S MARITIME BOUNDARIES

## *Arif Havas Oegroseno*

Indonesia is the largest archipelagic state in the world. Its vast maritime area of 5.8 million square kilometres consists of 0.3 million square kilometres of territorial sea, 2.8 million square kilometres of archipelagic waters and 2.7 million square kilometres of exclusive economic zone (EEZ). The country occupies a particularly important strategic location at the crossroads of two great oceans, the Indian Ocean and the Pacific Ocean, and two great continents, Asia and Australia. It links Europe, the Middle East and Africa with the world economic powerhouses in Asia. Indonesia is also the home of three major maritime 'choke' points: the Malacca and Singapore straits, the Sunda Strait and the Lombok Strait. These are not international straits, but straits under national sovereignty used for international navigation supporting the transportation of high-value commodities such as oil, gas and industrial goods. They also serve as strategic routes for military vessels of countries with blue water navies.

For Indonesia, territorial and boundary issues, whether on land, at sea or even in the air, have always been a matter of national priority. In dealings with neighbouring countries, border issues have consistently been high on the agenda since Indonesia became an independent nation in 1945. In terms of national cohesiveness, one of the most pressing territorial and border matters was the existence of high seas between Indonesia's islands, making Indonesia a collection of island enclaves in the midst of the high seas rather than an island nation unified by the sea.

## THE JUANDA DECLARATION AND INDONESIA'S BASELINES

To achieve its ideal of a nation unified, rather than separated, by the sea, on 13 December 1957 the Indonesian government under Prime Minister Juanda proclaimed that all waters between the islands of Indonesia came under Indonesian sovereignty. The Juanda Declaration, as it came to be known, was a challenge to the law of the sea as it was then applied. The concept of the archipelagic state as a fundamental principle of international law only gained international acceptance with the adoption and entry into force of the 1982 United Nations Convention on the Law of the Sea (UNCLOS). Through this convention, Indonesia gained legal recognition of its status as a archipelagic state. The formal acknowledgment that it exercised sovereignty over the waters between its islands proved a powerful unifying force, binding the Indonesian nation and state.

Indonesia's success in having the archipelagic principle recognized under international law constituted a major victory for Indonesian diplomacy. It showed that a relatively young nation could provide leadership to the international community and persuade it to change the law of the sea. Through the change in prevailing international law, Indonesia gained vast additional territory, not by military force but through diplomacy. The entry into force of UNCLOS in 1986 could even be regarded as a second declaration of Indonesia's independent existence, consigning to history the previous international regime under which the waters between Indonesian islands were treated as high seas.

The Juanda Declaration became a legal instrument in 1960 with the passage of Law No. 4/1960 on Indonesian Waters. This critical law defined the archipelagic baselines within which all waters would be designated Indonesian archipelagic waters. The baselines were to connect the outermost points of the outermost islands of the Indonesian archipelago. After a thorough survey to ascertain the location of the points used to define the ends of the baselines (as required by Law No. 4/1960), on 28 June 2002 the government adopted Government Regulation No. 38/2002 listing the exact coordinates of its archipelagic basepoints.

Since then, two major developments have necessitated the revision of Government Regulation No. 38/2002: the creation of the independent state of Timor-Leste on 20 May 2002; and the judgment of the International Court of Justice on 17 December 2002 that the islands of Sipadan and Ligitan belonged to Malaysia. In response to these developments, the Indonesian government issued a revised list of coordinates on 19 May 2008, set out in Government Regulation No. 37/2008 amending Government Regulation No. 38/2002. The text of both regulations was deposited with the United Nations in 2008. Indonesia's current baselines are shown in Map 2.1 (see page 29).

## INDONESIA'S MARITIME ZONES

Under UNCLOS 1982 and the two Indonesian government regulations, Indonesia has differing levels of sovereignty and jurisdiction over the waters within and outside its archipelagic baselines. Waters within the baselines comprise internal waters and archipelagic waters. Waters outside the baselines comprise territorial waters, a contiguous zone and an EEZ. Indonesia also has sovereignty and jurisdiction over the seabed and subsoil of the continental shelf beyond its archipelagic baselines. In accordance with article 48 of UNCLOS 1982, Indonesia projects its territorial waters, contiguous zone, EEZ and continental shelf from the archipelagic baselines designated in the two government regulations.

## INTERNAL WATERS

Article 8 of UNCLOS 1982 defines internal waters in the following manner: '... waters on the landward side of the baseline of the territorial sea form part of the internal waters of the State'. States have full sovereignty over these waters. Internal waters usually include such areas as ports, estuaries, inlets and the waters between islands. To date, Indonesia has not completed the task of charting the lines to demarcate its internal waters. This work is important, because separate legal regimes apply to internal waters and archipelagic waters.

### Archipelagic waters

Archipelagic waters are defined in article 49 of UNCLOS 1982. They comprise the waters enclosed by the archipelagic baselines of an archipelagic state, regardless of their depth or distance from the coast. The sovereignty of an archipelagic state extends to the air space above the archipelagic waters as well as to the seabed and subsoil and the resources they contain.

As Hasjim Djalal points out in Chapter 4 of this volume, the creation of archipelagic waters as a legal category was accompanied by the requirement to designate archipelagic sea lanes through those waters. Indonesia commenced this task in 1994 by conducting national surveys and interagency coordination. In 1996, Indonesia held talks with the International Hydrographic Organization and began consultations with interested user states, such as Australia, Japan, the United States and the United Kingdom. The government also commenced work on a submission to the International Maritime Organization (IMO) on the designation of the archipelagic sea lane passages. In 1998, the IMO endorsed Indonesia's proposal for three north–south archipelagic sea lanes (see Map 2.1). But with the creation of an independent Timor-Leste and the loss of the islands of Sipadan and Ligitan to Malaysia, the course of the eastern and central sea lanes (II and IIIB in Map 2.1) is now subject to change.

## Territorial seas

Indonesia has a vast area of territorial waters. In general, these waters stretch 12 nautical miles from the normal and straight baselines laid down by UNCLOS 1982.[1] In the narrower sections of the Malacca and Singapore straits, Indonesia's territorial waters are less than 12 nautical miles wide, because of the proximity of neighbouring states.

UNCLOS 1982 sets out the rights of coastal states in their territorial waters. These rights have been incorporated into Indonesian law through Government Regulation 36/2002, which allows foreign vessels the right of innocent passage through Indonesian territorial waters. Under international law, Indonesia may temporarily suspend the right of innocent passage through its territorial seas in certain circumstances.

## Contiguous zone

The definition of a contiguous zone is set out in article 33 of UNCLOS 1982. A contiguous zone may not extend further than 24 nautical miles from the territorial sea baseline. Where a coastal state claims a contiguous zone, it may exercise the control necessary to (a) prevent infringements of its customs, fiscal, immigration or sanitary laws and regulations within its territory or territorial sea; and (b) punish infringements of the above laws and regulations committed within its territory or territorial sea. Indonesia has a vast band of contiguous zones extending the full 24 nautical miles from its territorial sea baselines. Only in a narrow section of the Malacca Strait, and possibly in the Sulawesi Sea near the Indonesian islands of Miangas and Marampit, is the extent of Indonesia's contiguous zone limited by other jurisdictions. Indonesia has not issued specific national regulations on its contiguous zones but a draft regulation on this matter has been discussed.

## Exclusive economic zone

A coastal state has the right to claim an EEZ extending 200 nautical miles out from its territorial sea baselines. The rights, jurisdiction and duties of a coastal state in its EEZ are set out in article 56 of UNCLOS 1982. Among other things, it gives the state:

> sovereign rights for the purpose of exploring and exploiting, conserving and managing the natural resources, whether living or non-living, of the waters superjacent to the seabed and of the seabed and its subsoil, and with regard

---

1   See Chapter 5 by Arsana and Schofield (page 72) for an explanation of the differences between normal and straight baselines.

to other activities for the economic exploitation and exploration of the zone, such as the production of energy from the water, currents and winds.

Indonesia's claims to an EEZ are set out in Law No. 5/1983 on the Indonesian Exclusive Economic Zone, enacted in 1983. In some areas, such as the seas to the west of Sumatra, Indonesia has been able to claim the maximum breadth of 200 nautical miles from its territorial sea baseline. In others, the area it can claim is restricted by the overlapping claims of neighbouring countries and must be settled by international agreement. Indonesia has reached agreement with Australia and Papua New Guinea on the boundaries of overlapping EEZs, but has yet to reach agreement with other countries.

## Continental shelf

Article 76 of UNCLOS 1982 defines the limits of the continental shelf, and the extent to which a coastal state can claim jurisdiction over a continental shelf that extends beyond 200 nautical miles from its coastline. Article 77 details the rights of a coastal state over its continental shelf. Article 78 states that the rights of a coastal state over its continental shelf do not affect the legal status of the waters and airspace above it; it explicitly states that such rights 'must not infringe or result in any unjustifiable interference with navigation and freedoms of other States' as provided for elsewhere in the convention. Rather oddly, Indonesia's law on the continental shelf, Law No. 1/1973, still refers to the 1958 (rather than 1982) Convention on the Law of the Sea, because it has not been updated to reflect the changes that have taken place since it was enacted in 1973. Most of Indonesia's continental shelf has been delimited. The remaining areas lie in the Sulawesi Sea, the Pacific Ocean and the waters adjacent to Timor-Leste.

## Extended continental shelf

Indonesia is one of the states that may be able to claim continental shelf rights beyond 200 nautical miles based on article 76(4) of UNCLOS 1982. Indonesia believes it has a solid claim to an extended continental shelf in three areas: one to the northwest of Sumatra, one to the south of Sumba and one to the north of Papua. As discussed in more detail by Arsana and Schofield (see Chapter 5), Indonesia has lodged a submission with the United Nations Commission on the Limits of the Continental Shelf setting out its claim to an area to the northwest of Aceh. The commission will hear Indonesia's presentation in May 2009. Indonesia has started informal consultations with Papua New Guinea on the area to the north of Papua, with a view to collaborating on a joint submission. There is also the possibility of a trilateral submission with Palau, if it can be persuaded to cooperate.

## INDONESIA'S MARITIME BOUNDARIES

Indonesia has maritime boundaries with 10 countries, more than any other country in the world. Those countries are India, Thailand, Malaysia, Singapore, Vietnam, the Philippines, Palau, Papua New Guinea, Australia and Timor-Leste. In some areas, the geographic configuration means that boundaries would need to be negotiated trilaterally, not just bilaterally. These boundaries relate variously to Indonesia's territorial waters, contiguous zones, EEZ and continental shelf. So far, Indonesia has concluded treaties with all neighbouring countries except Timor-Leste and Palau. Treaties covering all shared maritime boundaries have been concluded with just two countries, Australia and Papua New Guinea.

### Indonesia–India

Three maritime boundary agreements have been signed by Indonesia and India. The first was the Agreement relating to the Delimitation of the Continental Shelf Boundary in the Great Channel between Sumatra and Nicobar Island (Andaman Sea), signed in Jakarta on 8 August 1974. This agreement was ratified by Presidential Decree No. 31/1974. The second was the Agreement relating to the Extension of the 1974 Continental Shelf Boundary in the Andaman Sea and the Indian Ocean. It was signed in New Delhi on 14 January 1977 and ratified by Presidential Decree No. 26/1977. The third was a trilateral agreement between Indonesia, India and Thailand on the determination of the junction point and demarcation of the boundaries of the three countries in the Andaman Sea. It was signed in New Delhi on 22 June 1978 and ratified by Presidential Decree No. 24/1978.

### Indonesia–Thailand

Indonesia and Thailand have concluded two bilateral maritime boundary agreements in addition to the trilateral agreement mentioned above. They are the Agreement relating to the Delimitation of a Continental Shelf Boundary in the Northern Part of the Straits of Malacca and in the Andaman Sea, signed in Bangkok on 17 December 1971 and ratified by Presidential Decree No. 21/1972; and the Agreement relating to the Delimitation of the Seabed Boundary in the Andaman Sea, signed in Jakarta on 11 December 1975 and ratified by Presidential Decree No. 1/1977.

## Indonesia-Malaysia

Indonesia and Malaysia have concluded three agreements related to the demarcation of their maritime boundaries. The first was an agreement relating to the delimitation of the continental shelf boundaries in the Malacca Strait and Natuna Sea. It was signed in Kuala Lumpur on 27 October 1969 and ratified by Presidential Decree No. 86/1969. The second concerned the delimitation of the territorial sea in the Malacca Strait. It was signed in Kuala Lumpur on 17 March 1970 and ratified by Law No. 2/1971. The third was a trilateral agreement between Indonesia, Malaysia and Thailand on continental shelf boundaries in the northern part of the Malacca Strait. It was signed in Kuala Lumpur on 21 December 1971 and ratified by Presidential Decree No. 20/1972.

## Indonesia-Singapore

Indonesia and Singapore have signed one maritime boundary agreement relating to the delimitation of the territorial sea in the Singapore Strait. It was signed in Jakarta on 25 May 1973 and ratified by Law No. 77/1973.

## Indonesia-Vietnam

Indonesia has signed one agreement with Vietnam relating to the delimitation of the continental shelf boundary. This agreement was signed in Hanoi on 26 June 2003 and ratified by Law No. 18/2007. The negotiations leading up to the treaty were very prolonged; the issue was first discussed as far back as 1978.

## Indonesia-Papua New Guinea

Indonesia and Papua New Guinea have concluded two maritime boundary agreements. The first, concerning boundaries in the Pacific Ocean, was signed between Indonesia and Australia on 12 February 1973 and ratified by Law No. 6/1973. The second, between Indonesia and Papua New Guinea, concerned the maritime boundaries of the continental shelf and EEZ in the Pacific Ocean. It was signed in Jakarta on 13 December 1980 and ratified by Presidential Decree No. 21/1982.

## Indonesia-Australia

Indonesia and Australia have concluded three agreements on maritime boundaries. The first, the Agreement Establishing Certain Seabed Boundaries, established the eastern segment of the seabed boundary between Indonesia and Australia in the Arafura Sea, as well as Indonesia's maritime boundary with Papua New Guinea in the Torres Strait. It was signed in Canberra on 18 May 1971 and ratified by Presidential Decree No. 42/1971. The second was a supplementary agreement covering the western segment of the seabed boundary in the Arafura Sea and part of the seabed boundary in the Timor Sea. It was signed in Jakarta on 9 October 1972 and ratified by Presidential Decree No. 66/1972. The third, the Agreement Establishing an Exclusive Economic Zone Boundary and Certain Seabed Boundaries, demarcated a long maritime boundary running from the Arafura Sea to the Indian Ocean south of Java. While accepting Australia's rights over the seabed of an extended continental shelf in the Timor Sea, it recognized Indonesia's ownership of resources in the water column of an EEZ based largely on the coordinates of the 1981 Provisional Fisheries Surveillance and Enforcement Line.[2] This agreement was signed on 14 March 1997 but has not been ratified.

## BORDER POLICY AND PRIORITIES FOR MARITIME BOUNDARY NEGOTIATION

From the Andaman Sea and Malacca Strait in the north to the Timor Sea in the south, a number of maritime boundaries between Indonesia and neighbouring countries have yet to be negotiated. Table 3.1 summarizes these boundaries by type and location.

In negotiating the country's boundaries, the Indonesian government places great importance on border diplomacy. It has established bilateral border commissions to facilitate the demarcation of boundaries and has cooperated extensively with neighbouring states on other border-related matters. Issues relating to land or maritime demarcation are important elements in border diplomacy, and are always present in bilateral deliberations with neighbours. In a clear demonstration of the importance of maritime boundary delimitation in Indonesian foreign policy, in 2007 Indonesia established diplomatic relations with Palau and opened negotiations on the two countries' maritime boundaries in the Pacific.

Indonesia established its first national negotiating team to handle maritime boundary negotiations in 1969. Initially led by Professor Mochtar Kusumaatmadja, the team continues to work with key institutions such as the Hydrographic Office, the National Coordinating Agency for Survey and Mapping, the army and navy, and the departments of Foreign

---

2  The Provisional Fisheries Surveillance and Enforcement Line was established to resolve a dispute over overlapping jurisdictions in the Timor Sea. See Chapter 12 by Fox (page 199) for further details.

Table 3.1    Summary of maritime boundaries still to be demarcated

| Neighbouring country | Type of boundary | Location of boundary |
|---|---|---|
| India | EEZ | Andaman Sea |
| Thailand | EEZ | Andaman Sea |
| India, Thailand | Tri-point of EEZ | Andaman Sea |
| Thailand, Malaysia | Tri-point of EEZ | Northern part of Malacca Strait |
| Malaysia | EEZ, territorial sea | Malacca Strait |
| Malaysia, Singapore | Tri-point of territorial sea | Singapore Strait |
| Singapore | Territorial sea | Western and eastern part of Singapore Strait |
| Singapore, Malaysia | Tri-point of territorial sea | Singapore Strait |
| Malaysia | Territorial sea | Singapore Strait |
| Malaysia | EEZ | Natuna Sea |
| Vietnam | EEZ | Natuna Sea |
| Malaysia | Territorial sea, EEZ, continental shelf | Sulawesi Sea |
| Philippines | Territorial sea, EEZ, continental shelf | Sulawesi Sea |
| Palau | EEZ, continental shelf | Pacific Ocean |
| Timor-Leste | Territorial sea, EEZ, continental shelf | Timor Sea, Sawu Sea |

Affairs, Defence, Energy and Mineral Resources, and Marine Affairs and Fisheries.

Unlike the original team, today's team is able to benefit from the advice and expertise of a highly qualified advisory board. The current members of the board are Ambassador Professor Hasjim Djalal, Ambassador Nugroho Wisnumurti (also an International Law Commission member), Professor Eti Agoes (University of Padjadjaran) and Professor Hikmahanto Juwana (University of Indonesia); the late Ali Alatas, a former foreign minister and ambassador to the United Nations, was also a member. It is their task to help the negotiating team formulate a national position. Indonesia also retains international counsel from time to time.

The normal procedure for formulating a negotiating position begins with the preparation of thorough legal and technical studies. These are presented to the advisory board, which prepares recommendations. Ministerial instruction is sought to complete the basic national position. The negotiating team then takes the agreed position to the negotiating table. The results of each round of negotiations are analysed thoroughly by the team and the advisory board. The board reports progress to the minister and awaits further instructions.

The science of maritime delimitation has developed significantly since the entry into force of the 1982 Convention on the Law of the Sea. Indonesia's negotiators are equipped with the latest chart and mapping technology and given the legal resources to analyse the numerous third-party adjudications that may be relevant to their maritime boundary negotiations. This is supplemented by legal and technical training for members of the negotiating team and officials in the relevant government agencies, conducted regularly by international experts.

# 4 INDONESIA'S ARCHIPELAGIC SEA LANES

*Hasjim Djalal*

---

At the time of independence, the Indonesian state inherited all former Dutch territories in the Indonesian archipelago, from the island of Rondo to the northwest of Aceh to Merauke in Papua (see Map 1.1 on page 2). With this territory, Indonesia also inherited, in accordance with international law at the time, three miles of territorial waters around each island. As the Dutch did not have sovereignty over the waters between those islands, Indonesia did not acquire jurisdiction over those waters when it became independent. Rather, the old colonial boundaries were preserved in the construction of the post-colonial state by virtue of the principle of *uti possidetis juris* (Latin: 'as you possess in law').

Instead of uniting the nation, the waters between Indonesia's islands became a barrier to the unity and integrity of the state. The status of the waters within the archipelago was governed by the regime of international waters or high seas, even though they were used on a daily basis for domestic travel within Indonesia. More serious from a security point of view was the fact that Indonesia consisted of pockets of sovereignty separated by international waters that could be exploited by other nations for purposes ranging from economic exploitation to the achievement of security objectives.

To overcome this anomaly, on 13 December 1957 Indonesia proclaimed itself an archipelagic state (*negara kepulauan*) by virtue of a declaration by Prime Minister Juanda Kartawijaya. The Juanda Declaration was later enacted through Law No. 4/1960 on Indonesian Waters. This declaration gave rise to one of the most difficult and contentious issues discussed between 1973 and 1982 at the United Nations Conference on the Law of the Sea, namely the status of the waters between the islands that constitute archipelagic states such as Indonesia. The final result of a painstaking and long process of negotiation was the acceptance of the principle of the archipelagic state as outlined in the 1982 United Nations Convention on the Law of the Sea (UNCLOS), the convention that gave birth to a new legal regime at sea. The entire fourth chapter of the convention was dedicated to the question of archipelagic states (articles 46–54). Article 49 recognized the political and territorial unity of an archipelagic state, stipulating that the sovereignty of an archipelagic state extended to the waters enclosed by its straight archipelagic baselines, as well as the airspace, seabed, subsoil and all resources contained therein, including in the seabed and subsoil as well as in the water column. Thus, in 1982 Indonesia's status as a single archipelagic state was recognized internationally under UNCLOS. Since that time, the sovereignty of Indonesia over its archipelagic waters, including the airspace above those waters and the seabed and subsoil below them, together with all the resources therein, has been guaranteed by international law.

## INTERNATIONAL SHIPPING AND ARCHIPELAGIC STATES

One of the key issues discussed at the Conference on the Law of the Sea was how to recognize the interest of archipelagic states in maintaining unity and exercising sovereignty over their archipelagic waters while at the same time guaranteeing other states passage through those archipelagic waters. Accordingly, the 1982 convention recognized the waters between the islands of an archipelagic state — previously regarded as international waters — as the sovereign territory of that archipelagic state, but also explicitly guaranteed foreign ships the right of transit through archipelagic waters. The means by which these two principles were to be reconciled was the designation of archipelagic sea lanes for the exercise of passage through archipelagic waters.

The convention reaffirmed the existing right of all ships to what is known as innocent passage through the territorial waters of another state. 'Innocent passage' is defined in some detail in the convention but it amounts to the right to pass promptly through a country's territorial waters, doing nothing that is not directly related to that passage. Trading, fishing, surveying and military display are therefore all excluded from the understanding of innocent passage. Under UNCLOS, this right of innocent passage through territorial waters applies equally to archipelagic waters. In addition, however, the convention provides for archipelagic sea lane passage — similar to the principle of transit passage through straits used for international navigation — within designated archipelagic sea lanes.

The 1982 convention also caters specifically for the interests of countries adjacent to archipelagic states that may be affected by the existence of archipelagic waters. Under the convention, an archipelagic state is required to respect existing agreements with other states, permit the maintenance and replacement of existing underwater cables upon certain conditions, and recognize traditional fishing rights and other legitimate activities of the immediately adjacent neighbouring states in accordance with specific bilateral agreements or arrangements between the archipelagic state and the neighbouring state concerned. The rules governing archipelagic sea lanes do not prejudice the right of foreign vessels to innocent passage outside those lanes.

In other words, there are two kinds of rights of navigation through Indonesian waters, the right of passage through designated archipelagic sea lanes and the right of innocent passage outside them. In broad terms, there are fewer restrictions on vessels using archipelagic sea lane passage than on those using the right of innocent passage. Under the principle of innocent passage, submarines and other underwater vehicles are required to navigate on the surface, showing their flags, but in archipelagic sea lanes, vessels are allowed to navigate in their 'normal mode', which allows for the possibility of underwater passage. In areas where innocent passage is permitted, there is no right of overflight, but flight over archipelagic sea lanes is permitted. The right of innocent passage though territorial waters can be suspended (for instance, to permit military exercises or to deal with local crises), whereas archipelagic sea lane passage cannot be suspended, although the sea lanes themselves can be substituted. Similarly, there are no precise rules under international law that would require countries to provide prior notification or prior authorization for warships navigating territorial seas under the right of innocent passage. Some countries require prior notification, some require prior authorization and some strongly oppose such requirements. After years of debate at the third United Nations Conference on the Law of the Sea (LOSC III), UNCLOS remains silent on this issue. Under the right of archipelagic sea lane passage, in contrast, there is explicitly no requirement for prior notification or prior authorization for the passage of warships, although it is recommended that they inform the archipelagic state concerned before undertaking passage through a designated sea lane.

Another matter that is not covered by the principles of archipelagic sea lane passage is the possibility of cooperation between user states and archipelagic states, firstly to establish and maintain the necessary navigational and safety aids to support international navigation, and secondly to prevent, reduce and control pollution from ships. Such a provision does exist in relation to the right of transit passage through straits used for international navigation (UNCLOS, article 43). Originally, archipelagic states argued that the provision would reduce the capacity of an archipelagic state to exercise sovereignty over its archipelagic waters. Moreover, this provision was regarded by some as important in differentiating between navigation in straits and navigation in sea lanes. However, since cooperation between user and archipelagic states in the promotion of navigational safety and the prevention of pollution in archipelagic waters is not prohibited by UNCLOS, such cooperation could be applied to archipelagic waters if the archipelagic state concerned so decided.

## CREATING INDONESIA'S SEA LANES

For Indonesia, achieving acceptance of the archipelagic concept as an international legal principle involved a trade-off, in that it had to recognize a right of passage through Indonesian waters which went beyond the normal right of innocent passage. The outcome of this trade-off was the designation of archipelagic sea lanes, described in article 53 of the 1982 convention. The designation of sea lanes for the exercise of archipelagic sea lane passage is not mandatory, since the operative word 'may' is included in article 53(1) of UNCLOS. The catch, from Indonesia's point of view, comes in articles 53(4) and 53(12), under which a state that chooses to designate any archipelagic sea lanes must designate lanes to cover 'all normal passage routes used as routes for international navigation or overflight'. Until this is done, the right of archipelagic passage applies by default to all routes 'normally used for international navigation'. In other words, in practice Indonesia can only make full use of its sovereignty over its archipelagic waters under the archipelagic principle if it designates a full regime of archipelagic sea lanes through its seas. Indonesia's strategic location between two continents, Asia and Australia, and two great oceans, the Indian Ocean and the Pacific Ocean, and the fact that a number of strategic straits are located in Indonesia mean that sea lanes must be established to facilitate passage through those straits. The contentious issue now is how numerous those archipelagic sea lanes should be and where they should run.

The process of designating Indonesia's archipelagic sea lanes has been undertaken in a number of stages, involving surveys, national coordination meetings, and consultations with neighbouring states and other interested parties, particularly Australia, the United States and the International Hydrographic Organization. Indonesia commenced the process of designating archipelagic sea lane passages by conducting a series of national surveys in 1994, and completed the requisite national interagency coordination in 1995. In 1996, Indonesia held consultations with the International Hydrographic Organization and began consultations with interested user states, such as Australia, Japan, the United States and the United Kingdom, on the designation of the archipelagic sea lanes and the rules that would govern their use. These consultations culminated in a general agreement or understanding on 19 rules that would be applicable in the sea lanes (see Box 4.1). In the same year, Indonesia began assembling a submission to the International Maritime Organization (IMO) on the designation of its archipelagic sea lane passages.

The Indonesian submission came under consideration at the 67th Meeting of the Maritime Safety Committee of the IMO in 1998, where the IMO accepted the partial designation of Indonesian archipelagic sea lane passages. The sea lanes adopted at that meeting (under resolution 72(69)) consisted of three routes – Archipelagic Sea Lanes I, II and III – running along a north–south axis (see Map 2.1 on page 29). The adoption of these three archipelagic sea lanes was later enacted in Indonesia under Government Regulation No. 37/2002 dated 28 June 2002. Archipelagic Sea Lane I had two branches in the Karimata Strait, intended to facilitate navigation from the Indian Ocean through the Sunda Strait to the Natuna Sea and finally to the South China Sea. Archipelagic Sea Lane II was designed to facilitate navigation from the Indian Ocean through the Lombok Strait to the Makassar Strait and then to the Sulawesi Sea and the Pacific Ocean and Philippine waters. The branches of Archipelagic Sea Lane III served to facilitate navigation from the Timor Sea and the Arafura Sea to the Pacific Ocean through the Sawu Sea, the Banda Sea, the Seram Sea and the Molucca Sea.[1]

A number of factors were taken into consideration before the axis of the sea lanes was decided. These included technical requirements, such as the needs of international vessels and aircraft that wished to transit through or above Indonesian waters; the presence of maritime installations and structures, including underwater cables and pipelines; the capacity of law enforcement agencies to monitor navigation and overflight in the relevant areas so that law and order could be safeguarded; and the need to maintain peace, stability and security in Indonesia's heavily populated coastal zones. Other factors concerned the natural environment, including hydrographic and natural marine conditions in and near the relevant axis lines; the intensity of coastal and interisland navigation and overflight; the intensity of fishing activities, particularly of local artisanal fishermen; the existence of oil and gas exploration and exploitation activity; the need to protect the marine environment, marine

---

1  East Timor (Timor-Leste) gained independence in 1998, after these three archipelagic sea lane passages were adopted. This political development may require an adjustment of the designated archipelagic sea lanes at some future date.

## Box 4.1    Sea lane rules

1   Ships and aircraft in sea lanes will not disturb or threaten the sovereignty, territorial integrity or independence and national unity of Indonesia. They will not carry out any action that would contravene principles of international law as embodied in the UN Charter.

2   Except for situations involving force majeure or distress, aircraft in archipelagic sea lane passages shall not land in Indonesian territory, including territory in the sea lanes. Ships and aircraft in archipelagic sea lane passages shall not deviate more than 25 nautical miles to either side of axis lines during passage, provided that such ships and aircraft shall not navigate closer to the coast than 10 per cent of the distance between the nearest points on the islands bordering the sea lanes.

3   Foreign civil aircraft passing through the sea lanes must comply with international rules of civil aviation as established by the International Civil Aviation Organization (ICAO).

4   While exercising sea lane passage, foreign warships and foreign military aircraft are not allowed to conduct war exercises or use live ammunition or conduct war games. They are to proceed without delay through or over the sea lanes in the normal mode, solely for the purpose of continuous, expeditious and unobstructed transit.

5   Foreign warships and military aircraft as well as ships using nuclear energy that pass through the sea lanes are advised to inform the Indonesian government (the Commander of the Indonesian Armed Forces) in advance for the purpose of safety of navigation, and to take the necessary preparatory actions to prevent something untoward happening.

6   Ships carrying nuclear material, except warships and other government ships operated for non-commercial purposes, are required to notify the Commander of the Indonesian Armed Forces in advance in accordance with the Convention on the Physical Protection of Nuclear Materials; are requested to comply with the International Code for the Safe Carriage of Packaged Irradiated Nuclear Fuel, Plutonium and High-level Radioactive Wastes on Board Ships (the INF Code); and are required to comply with other international conventions dealing with the transportation or carriage of dangerous goods, hazardous materials and noxious substances.

7   Foreign military aircraft flying above the sea lanes must observe the safety of civil aviation, monitor the emergency frequencies and maintain contact with the authorized air traffic controllers.

8   Transiting foreign ships and/or aircraft should move carefully in sea lanes where there is abundant economic activity (fisheries or mining), exercise caution in limited navigational areas, not come within 500 metres of an oil or gas installation, and observe the location of and be careful with respect to underwater cables and pipelines.

9   Foreign fishing vessels must keep their fishing gear stowed during transit, and are prohibited from carrying out fishing activities while in transit.

10  Ships transiting through sea lanes must follow generally accepted international navigational rules for the safety of navigation and show due regard for local shipping as well as the activities of local fishermen.

11  Ships in sea lane passages shall comply with all generally accepted international standards regulating pollution of the marine environment from vessels, and shall not dump waste or other matter or discharge poisonous or dangerous materials while in Indonesian waters.

12  All ships are prohibited from cleaning their tanks or polluting Indonesian waters while in transit.

13  Ships in sea lane passages are not allowed to stop or anchor or move back and forth without a legitimate reason, except in cases of force majeure or distress. Transiting ships shall navigate in the normal mode solely for the purpose of continuous, expeditious and unobstructed transit.

14  Transiting ships are not allowed to disembark persons or goods and transfer them to other ships in contravention of the customs, financial, immigration or health rules of Indonesia, or carry out other activities in contravention of those rules.

15  Transiting ships and aircraft are not allowed to carry out survey work or marine scientific research, including taking water samples, for the purpose of investigation during passage. They shall not interfere with the survey or marine scientific research activities carried out by Indonesia in the sea lanes.

16  Transiting ships and aircraft are prohibited from carrying out unauthorized broadcasting or emitting electromagnetic signals that are aimed at interfering with national telecommunications systems, and are prohibited from establishing direct communications with unauthorized persons or certain groups in the territory of Indonesia.

17  Transiting ships shall always meet the generally accepted international requirements for the safety of navigation.

18  To the maximum extent established by applicable international agreements, shippers, cargo owners and ship owners are strictly liable, individually or collectively, for the damage caused by them, including paying compensation to Indonesia, by way of applicable international agreements or otherwise. Accordingly, they shall be insured as required. Such liability and insurance shall include compensation for damage, including direct, consequential or environmental damage, to Indonesia. The flag state of a ship entitled to sovereign immunity shall bear international responsibility for damages caused to Indonesia by the non-compliance of the ship with international law.

19  To ensure the safety of navigation and the safety of Indonesia, it is recommended that foreign tankers, vessels using nuclear energy, foreign vessels carrying nuclear substances and other dangerous goods, foreign fishing vessels and foreign warships passing through Indonesian waters from one part of the exclusive economic zone or high seas to another part of the exclusive economic zone or high seas use the sea lanes.

parks and marine ecosystems; and the extent of and potential for coastal and marine tourism.

Government Regulation No. 37/2002 was based closely on the rules of UNCLOS 1982 as well as on subsequent consultations with the various users of the archipelagic sea lanes, particularly military vessels and aircraft. It affirms that all foreign vessels have the right of innocent passage through Indonesian territorial seas and archipelagic waters, without entering internal waters, for the purpose of transiting from one part of the exclusive economic zone or high seas to another part of the exclusive economic zone or high seas (article 2(1)). It also states that foreign vessels and aircraft can exercise the right of archipelagic sea lane passage through certain parts of the Indonesian territorial seas and archipelagic waters, namely the specific archipelagic sea and air routes stipulated in the regulation (articles 2, 3 and 11). The regulation specifies that passage must be carried out quickly in the 'normal mode' solely for the purpose of continuous, direct, expeditious passage and unobstructed transit (article 4(1); UNCLOS, article 53(3)).

Under the regulation, ships and aircraft exercising the right of archipelagic sea lane passage may not deviate more than 25 nautical miles from the axis of the archipelagic sea lane and may not navigate or fly closer to the coast than 10 per cent of the distance between the nearest points on the islands bordering the sea lane (article 4(5); UNCLOS, article 53(5)). The regulation also specifies that during passage vessels may not threaten or use force against the sovereignty, territorial integrity or political independence of Indonesia or in any other manner violate the principles of international law as embodied in the Charter of the United Nations (article 4(3); UNCLOS, articles 39(1b) and 34). More specifically, in exercising the right of archipelagic sea lane passage, foreign vessels or aircraft may not carry out war exercises or exercises that involve weapons with ammunition (article 4(4)). The flag state of the vessels or country of registry of the aircraft shall bear international responsibility for any loss or damage suffered by Indonesia as a result of non-compliance with these rules by warships or aircraft.

Foreign vessels in an archipelagic sea lane passage must follow the generally accepted international rules, procedures and practices on safety of navigation, including rules on avoiding collision at sea (article 7(3)), and must follow the Traffic Separation Scheme to ensure safety of navigation (article 7(2)). Foreign aircraft in an archipelagic sea lane passage must follow International Civil Aviation Organization (ICAO) rules (for civil aircraft), monitor the radio frequency indicated by the competent authority (or the relevant international emergency radio frequency) (article 8) and respect ICAO regulations on safety of aviation. Except in cases of force majeure or accident, foreign aircraft exercising their right of archipelagic sea lane passage may not land in Indonesian territory, and ships may not stop or anchor or zig-zag except to give assistance to people and ships in danger (articles 4(5) and 4(6)). In addition, foreign ships and aircraft in an archipelagic sea lane passage may not engage in unauthorized broadcasting (article 4(7); UNCLOS, articles 109 and 110), cause disturbance to telecommunications systems, communicate directly with unauthorized people or groups, or embark or disembark people, goods or currency in contravention of customs, immigration, fiscal and sanitary controls, except in cases of force majeure or distress (article 6(3)).

Foreign vessels or aircraft, including research or hydrographic survey vessels or aircraft, may not carry out maritime research activities or hydrographic surveys in an archipelagic sea lane passage, either by using detecting devices or by taking samples, unless given permission to do so (article 4(7); UNCLOS, article 40). Foreign fishing and other vessels may not carry out fishing activities, and foreign fishing vessels must keep their fishing equipment in storage (articles 6(1) and 6(2)). Similarly, vessels may not cause disturbance or damage to navigational aids or facilities, or to underwater cables and pipelines (article 7(3)), and may not approach closer than 500 metres to an oil or gas installation or any other installation (article 7(4); UNCLOS, articles 60(5) and 80). They are also prohibited from discharging oil, oily waste and other pollutants into the marine environment, from carrying out activities contrary to international regulations and standards to prevent, reduce or control marine pollution from ships (article 9(1)) and from dumping in Indonesian waters (article 9(2)). Nuclear vessels or vessels carrying nuclear materials or other dangerous or hazardous materials in an archipelagic sea lane passage must carry the appropriate documentation and follow the special preventive measures prescribed by international agreement for such vessels (article 9(3)). Those responsible for the operation of commercial vessels or aircraft are liable and responsible for any loss or damage suffered by Indonesia as a result of non-compliance with the rules of archipelagic sea lane passage (article 10(1)).

## OUTSTANDING CONCERNS

Article 53(4) of UNCLOS implies that a state that chooses to designate any archipelagic sea lanes must designate them for all normal international shipping routes. In 1998, however, the IMO agreed to accept Indonesia's designation of three lanes as a partial designation. This agreement was reflected in article 3(2) of Government Regulation No. 37/2002, which allowed for de facto archipelagic sea lane passage in other parts of the archipelago even after the designation of the three north–south lanes.

This compromise arose as follows. During the consultations leading up to the adoption of the three north–south archipelagic sea lanes, Australia and the United States had proposed that east–west archipelagic sea lanes also be designated (see Map 2.1 on page 29). They maintained that the east–west passage constituted a 'normal' route under UNCLOS and that Indonesia was therefore obliged to designate archipelagic sea lanes to cover it. In July 2003, the United States asserted what it claimed to be its customary rights to east–west passage when it sent an aircraft carrier, the *Carl Vinson*, and five fighter aircraft on manoeuvres near the island of Bawean in the Java Sea. The incident caused consternation in Indonesia and led to a growing push domestically for Indonesia to regularize a single east–west route through the Java Sea. However, before any such lane is designated formally, thorough deliberation is required to ensure that navigational safety and the protection of the environment, as well as the security of Indonesia, would be safeguarded.

The Indonesian government has also noted that some user states claim to be navigating in international waters when passing through Indonesian archipelagic waters or using the archipelagic sea lanes even though, legally, archipelagic waters fall within the sovereignty of a coastal state and the right of archipelagic sea lane passage does not in any way affect that status (UNCLOS, article 49(4)). The government is unhappy with this interpretation but has so far tolerated breaches in the hope that a better understanding of the delicate balance that was so painstakingly negotiated over many years may develop. There is also the prospect of differing interpretations of the status of navigable straits around islands that lie wholly within 25 miles of the axis of an archipelagic sea lane passage, as in the case of the Sunda Strait. The axis line in the Sunda Strait lies to the west of Sangiang Island, but the waters to the east of the island are still within 25 miles of the line and are suitable for navigation. As a result, it is unclear whether the regime applies just to the western channel or to the channel east of the island as well.

Finally, there is a gap in international law with regard to the legal principle of airspace above archipelagic waters, because the 1944 Convention on International Civil Aviation (the Chicago Convention) has not been revised to reflect the current international law of the sea. Article 1 of the Chicago Convention stipulates that every state has complete and exclusive sovereignty over the airspace above its territory. However, article 2 states that the territory of a state shall be deemed to be its land areas and the territorial waters adjacent to them, presumably extending from three to 12 miles from the coastline. Clearly there is a need to revise the 1944 convention to reflect new developments in spatial international law.

The recognition of archipelagic states under UNCLOS was a major achievement for Indonesia. For the first time, Indonesia's sovereignty over the waters between its islands was recognized internationally. This new regime left important practical issues to be resolved, including the relationship between Indonesian sovereignty and the reasonable rights of international shipping to pass through Indonesian waters. The outcome was a new concept in international law, the archipelagic sea lane. Although the precise form of this new system has not been fully worked out, the designation of the most important of these sea lanes brings Indonesia one step closer to implementing its rights over, and accepting its responsibilities for, its archipelagic waters.

# 5 EXTENDING INDONESIA? OPPORTUNITIES AND CHALLENGES RELATED TO THE DEFINITION OF INDONESIA'S EXTENDED CONTINENTAL SHELF RIGHTS

*I Made Andi Arsana and Clive Schofield*

This chapter examines issues related to Indonesia's continental shelf and, in particular, those areas of continental shelf extending seaward of the 200-nautical-mile limit of Indonesia's exclusive economic zone (EEZ), which may form part of its legal continental shelf. Such areas are commonly referred to as the 'outer' or 'extended' continental shelf. The continental shelf can be considered to consist of the seabed and subsoil areas surrounding a coastal state's land territory. As a consequence of the link between a coastal state's land territory and the continental shelf surrounding it, coastal states possess sovereign rights over their continental shelf.

Under the United Nations Convention on the Law of the Sea (UNCLOS), in order to confirm their rights over extended continental shelf areas, eligible coastal states are required to make a submission to a specialized scientific body, the United Nations Commission on the Limits of the Continental Shelf. Indonesia is among the coastal states that are in a position to make such a submission and thereby confirm their sovereign rights over extended continental shelf areas.

On 16 June 2008, following approximately 10 years of preparatory work, Indonesia made a partial submission to the commission relating to an area to the northwest of Sumatra. Indonesia is understood to be in the process of preparing submissions in respect of two further areas of continental shelf beyond the 200-nautical-mile limit. Indonesia's submission was the first made by a Southeast Asian state and among the first on the part of a developing state. Confirmation of its rights over areas of continental shelf beyond 200 nautical miles from its baselines will offer Indonesia the opportunity to explore and exploit any natural resources discovered therein.

The deadline for submissions to the commission for many coastal states, including Indonesia, was 13 May 2009. However, demonstrating to the commission that seabed areas beyond the 200-nautical-mile limit properly form part of a coastal state's extended continental shelf is no easy task. Detailed geoscientific information is required in respect of the geology (composition) and morphology (shape) of the continental margins in question. Additionally, bathymetric (depth) data are needed, as are geodetically robust (that is, accurate) distance measurements from the state's coastal baselines. This necessitates the gathering of scientific information, notably through seismic and sonar surveys, and the interpretation of the data ahead of the submission, all of which is time consuming and expensive.

These challenges are illustrated through a preliminary assessment of Indonesia's partial submission. Indonesia has already invested considerably in making its partial submission to the commission and is continuing preparations which may lead to further 'extensions' of Indonesia. While the immediate benefits in economic terms may be hard to reconcile with the costs incurred in undertaking offshore surveys and preparing submissions, the cost–benefit equation may well change over time. More fundamentally, it is also the case that Indonesia's sovereign rights over an enlarged maritime area subject to national jurisdiction are at stake — something that is harder to quantify in purely economic terms.

## INDONESIA'S CLAIMS TO MARITIME JURISDICTION

The basic international legal framework governing maritime jurisdictional claims and the delimitation of maritime boundaries is provided by the 1982 United Nations Convention on the Law of the Sea, which has gained widespread international acceptance, including ratification or accession by a majority of the world's coastal states.[1] Indonesia became

---

1  The convention was adopted in Montego Bay, Jamaica, on 10 December 1982 and came into force on 16 November 1994. As of 5 February 2009 there were 158 parties to the convention: 157 states plus the European Community. The text of the convention and related agreements are available at www.un.org/Depts/los/index.htm. A list of the current signatories is available at www.un.org/Depts/los/reference_files/status2008.pdf.

a party to the convention on 3 February 1986 (under Law No. 17/1985).[2] As other chapters in this book show, a number of Indonesia's maritime claims significantly pre-date UNCLOS and indeed these claims helped to inspire some of the provisions in the convention. In particular, Indonesia was one of the key pioneers of the archipelagic concept, having in 1960 unilaterally defined a system of straight baselines around the outermost Indonesian islands totalling over 8,000 kilometres in length (under Law No. 4/1960 on Indonesian Waters).[3] In so doing, Indonesia emphasized the unique 'characteristics and peculiarities' of Indonesia as an archipelagic state, and that the islands and waters lying between them 'should be regarded as a single unit' (Law No. 4/1960, article 1).

Baselines are fundamental to a coastal state's claims to maritime jurisdiction in that they provide the starting point for measuring its maritime claims offshore. According to UNCLOS (article 5), a coastal state's 'normal' baselines consist of the low-water line along the coast as marked on large-scale charts that it officially recognizes. The convention does, however, allow for a variety of straight-line types of baselines to be used as an alternative to normal baselines, where the relevant coastal geography merits such a departure.[4]

Indonesia's 1960 claims informed the drafting of article 47 of UNCLOS, which sets out the relevant provisions governing archipelagic baselines.[5] In 1996 Indonesia issued new legislation on Indonesian waters, Law No. 6/1996, relating to the definition of Indonesia's archipelagic baselines.[6]

---

2  Law No. 17/1985 was enacted on 31 December 1985. However, the United Nations recognizes 3 February 1986 as the relevant date for Indonesia, presumably because of a delay in delivering the notice of ratification of the law to the UN secretary-general.

3  Law No. 4/1960 is reproduced in US Department of State (1971).

4  UNCLOS provides for several types of straight-line baselines. These include straight baselines (article 7), river closing lines (article 9), bay closing lines (article 10), baselines in respect of ports (article 11) and roadsteads (waiting areas for shipping before entry into port) (article 12).

5  Article 46 sets out the requirements for archipelagic state status, while article 47 provides detailed rules relating to archipelagic baselines—notably that the claimant state's 'main islands' must be included within the archipelagic baseline system; that the ratio of water to land within the baselines must be between 1:1 and 9:1; that the length of any single baseline segment must not exceed 125 nautical miles; that no more than 3 per cent of the total number of baseline segments enclosing an archipelago may exceed 100 nautical miles; and that such baselines 'shall not depart to any appreciable extent from the general configuration of the archipelago' (see Tsamenyi, Schofield and Milligan 2008).

6  A translation of Law No. 6/1996 is available at http://www.un.org/Depts/los/LEGISLATIONANDTREATIES/PDFFILES/IDN_1996_Act.pdf.

The new law, which replaced Law No. 4/1960, defined Indonesia's archipelagic baselines in general terms. Supplementary regulations issued in 1998, 2002 and 2008 specifically designated (or redesignated) Indonesia's archipelagic baselines through lists of geographical coordinates.[7] The practical impact of the new law and associated regulations was to change some of Indonesia's archipelagic baselines, notably in the Natuna Sea and in the vicinity of Timor-Leste. Indonesia's current archipelagic baselines are shown in Map 2.1 (see page 29).

From its baselines Indonesia claims the full suite of zones of maritime jurisdiction provided for in accordance with UNCLOS. These zones include a territorial sea out to 12 nautical miles, a contiguous zone out to 24 nautical miles, an EEZ out to 200 nautical miles as well as the continental shelf (see Figure 5.1).[8]

In considering claims to maritime jurisdiction, it is important to distinguish between those zones within which coastal states, including Indonesia, exercise full sovereignty and those where they exercise only sovereign rights for specific purposes. Coastal states have sovereignty over internal waters (landward of defined baselines),[9] and over the territorial sea. Additionally, archipelagic states, such as Indonesia, have sovereignty over the waters within their archipelagic baselines, termed 'archipelagic waters'.[10] With regard to all these zones, the coastal state's

---

7   See Government Regulation (PP) No. 61/1998 on the List of Geographical Coordinates of the Base Points of the Archipelagic Baselines of Indonesia in the Natuna Sea (available at http://www.un.org/Depts/los/LEGISLATION ANDTREATIES/PDFFILES/IDN_1998_Regulation61.pdf); PP No. 38/2002 (available at http://www.dtic.mil/whs/directives/corres/20051m_062305/ indonesia.doc); and PP No. 37/2008 (available, in Indonesian, at http:// www.djpp.depkumham.go.id/inc/buka.php?czoyNDoiZD0yMDAwKzgm Zj1QUDM3LTIwMDguaHRtljs=).

8   Indonesia's territorial sea claims were reaffirmed through Law No. 6/1996. Indonesia made a claim to an EEZ through the Declaration by the Government of Indonesia on the Exclusive Economic Zone of Indonesia, 21 March 1980 (available at http://www.un.org/Depts/los/LEGISLATION ANDTREATIES/PDFFILES/IDN_1980_DeclarationEEZ.pdf), and through Law No. 5/1983 on the Indonesian Exclusive Economic Zone (available at http://www.un.org/Depts/los/LEGISLATIONANDTREATIES/PDF FILES/IDN_1983_Act.pdf).

9   Internal waters exist landward of defined straight-line types of baselines, such as the waters within a bay where a bay closing line has been defined or within a port where port closing lines have been constructed. Internal waters are considered to be under the full sovereignty of the coastal state, with no right of navigation for foreign states. This allows the coastal state, for example, to close its ports to foreign shipping if it so desires.

10  Article 46 of UNCLOS defines an 'archipelagic state' as a state 'constituted wholly by one or more archipelagos and may include other islands', and an

*Figure 5.1    Claims to maritime jurisdiction*

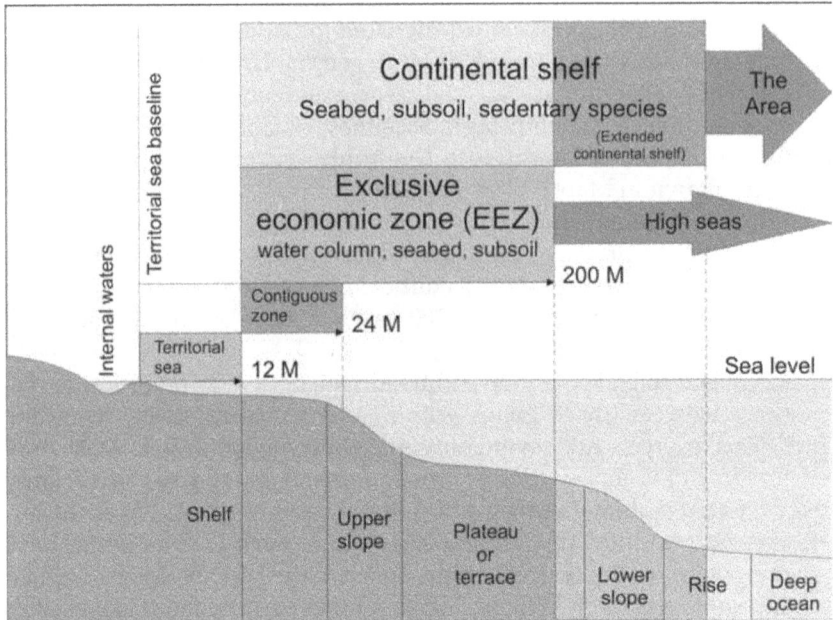

*Source:* Adapted from Geoscience Australia (2006).

sovereignty extends throughout the seabed and subsoil, water column and airspace above. It is important to note, however, that vessels of other states do retain some navigational rights within both territorial seas and archipelagic waters.[11]

Within the EEZ and continental shelf, in contrast, coastal states have specific sovereign rights that are largely focused on the utilization and management of natural resources (notably fisheries, oil and gas, and minerals). While the EEZ encompasses both the seabed (including sub-soil) and the water column overlying the seabed, the continental shelf

---

'archipelago' as 'a group of islands, including parts of islands, interconnecting waters and other natural features which are so closely interrelated that such islands, waters and other natural features form an intrinsic geographical, economic and political entity, or which historically have been regarded as such'. According to these definitions Indonesia is clearly an archipelagic as well as a coastal state.

11   Within territorial seas foreign vessels have the right of 'innocent passage' (UNCLOS, articles 17–26). UNCLOS also provides for both innocent pas-sage and 'archipelagic sea lanes passage' within archipelagic waters (articles 52–54) (see Tsamenyi, Schofield and Milligan 2008).

relates solely to the seabed and subsoil.[12] The high seas are those parts of the ocean that lie beyond national claims to internal waters, archipelagic waters (for an archipelagic state), territorial seas and EEZs (UNCLOS, article 86).[13] 'The Area', meanwhile, is the official but somewhat ambiguous name for that part of the seabed that is beyond national jurisdiction. While the limits of the Area will often coincide with those of the high seas — that is, at the 200-nautical-mile EEZ limit claimed by coastal states — this is not always the case because of the rights that certain coastal states may have over areas of extended continental shelf beyond the 200-nautical-mile limit. The Area is often also termed the 'international seabed' and is considered the common heritage of mankind (article 136).[14]

A further difference between the various maritime zones of jurisdiction is that some of them, such as the EEZ, require an active claim on the part of the coastal state, while others do not. Of particular relevance to the present discussion, continental shelf rights are inherent and 'do not depend on occupation, effective or notional, or on any express proclamation' (article 77(3)). This means that a coastal state may explore and exploit the natural resources on its continental shelf without making an express claim to it.

The basis for this distinct treatment rests on the concept that a coastal state possesses rights to its continental shelf as the 'natural prolongation of its land territory' into and under the sea — an idea well established prior to the conclusion of UNCLOS (article 76(1)). Following on from the proliferation of claims sparked by the so-called Truman Proclamation of 1945, when the United States asserted rights over the seabed seaward of its then three-nautical-mile territorial sea limit, coastal state rights over the continental shelf were enshrined in the 1958 Convention on the Continental Shelf.[15] Subsequently, in 1969, the International Court of Justice,

---

12  Part II of UNCLOS deals with the territorial sea and contiguous zone, Part V with the EEZ and Part VI with the continental shelf. It is worth noting, however, that rights over seabed and subsoil claimed as part of an EEZ under Part V of UNCLOS are exercised in accordance with Part VI (article 56(3)).

13  In this context, it is important to note that whereas coastal state rights within an EEZ are limited to specific sovereign rights, for instance over resources, certain freedoms of the high seas, notably freedom of navigation, are retained throughout the EEZs claimed by coastal states.

14  Exploitation of mineral resources in the Area is formally under the jurisdiction of the International Seabed Authority, an autonomous body with a formal relationship agreement with the United Nations (see footnote 43).

15  The convention was opened for signing on 29 April 1958 and came into force on 10 June 1964. This convention and three others (related respectively to territorial seas and contiguous zones, the high seas, and fishing and conservation of the living resources of the high seas) arose from the first United Nations Conference on the Law of the Sea, held in Geneva in 1958.

through its judgment in the North Sea continental shelf cases, stated that coastal states have rights over that part of the continental shelf that constitutes 'a natural prolongation of its land territory'. This concept, and the 1969 judgment of the court, made it clear that the continental shelf represents part of the inherent sovereign rights of a coastal state, and this informed the drafting of UNCLOS.

## INTRODUCING THE EXTENDED CONTINENTAL SHELF

Scientists have recognized since at least the early twentieth century that the surface of the earth is marked by a sharp distinction between continental and oceanic crusts. The continental crust is lighter and thicker, and much of it protrudes above the surface of the sea, thus creating the earth's continents. The outer margins of this crust are generally lower and are now submerged (though they have regularly been exposed during ice ages). This submerged portion of the continental crust is generally referred to as the continental shelf. In most places, the continental shelf ends abruptly, falling away to a relatively steep continental slope, ending in the gentler gradient of the continental rise, which in turn slopes down to the deep seabed or abyssal plane. In some cases the continental slope ends in a deep ocean trench. Depending on local conditions, therefore, the continental shelf may be broad or narrow. Java, Sumatra and Kalimantan are linked to the Asian mainland by the broad Sunda Shelf, which is covered by the Java Sea and part of the South China Sea. In contrast, there is hardly any continental shelf along the south coast of Java, where the land drops off quickly into the Java Trench. The continental shelf, continental slope and continental rise are together generally referred to as the continental margin.

As noted above, international law has long recognized the 'natural prolongation' of a coastal state's land territory into and under the sea to the continental shelf surrounding it, as a consequence of which coastal states possess sovereign rights over their continental shelf. Where the continental margin is broad and the continental shelf extends beyond 200 nautical miles from its baselines, a coastal state may possess rights over continental shelf areas seaward of its 200-nautical-mile limits — that is, over 'outer' or 'extended' continental shelf areas seaward of, for example, its EEZ.

Article 76 of UNCLOS defines the continental shelf of a coastal state as comprising 'the seabed and subsoil of the submarine areas' that extend beyond the limits of the state's territorial sea 'throughout the natural prolongation of its land territory to the outer edge of the continental margin' or, alternatively, to 200 nautical miles from its relevant baselines where the continental margin does not extend to that distance.

For many coastal states, therefore, the limit of the continental shelf is coincident with the 200-nautical-mile limit, because the continental margin on which the state is located is narrower than 200 nautical miles. Additionally, the configuration of coasts and islands and the proximity of other states mean that some coastal states are hemmed in by the maritime claims of other states and, in effect, 'shelf-locked'. However, where a coastal state is located on a broad continental margin that extends beyond the 200-nautical-mile limit, parts of that margin seaward of the limit may form part of the state's 'natural prolongation' and thus part of its continental shelf. In order for a coastal state to confirm its sovereign rights over areas of continental shelf beyond 200 nautical miles from its baselines, UNCLOS provides that it should make a submission regarding its proposed outer continental shelf limits to a specialized scientific body established under the convention, the United Nations Commission on the Limits of the Continental Shelf.[16] As noted above, the continental shelf beyond the 200-nautical-mile limit is often termed the 'outer' or 'extended' continental shelf. The latter term is somewhat misleading in that coastal states possess inherent continental shelf rights, so that Indonesia and other coastal states engaged in the same process are not 'extending' their maritime claims as such, but merely defining the outer limits of their continental shelf, thereby confirming their sovereign rights to continental shelf areas beyond their 200-nautical-mile limits.[17] Once the commission has delivered its recommendations to the coastal state, that state may declare the outer limits of its outer continental shelf to be 'final and binding' when defined 'on the basis of' the commission's recommendations (UNCLOS, article 76(8)).

Article 76 of UNCLOS provides a complex series of provisions according to which a coastal state can establish, largely on the basis of geological and geomorphologic evidence, the location of the outer limits of its extended continental shelf. The procedure for delineating the outer limits of a state's extended continental shelf is governed by article 76 of UNCLOS together with the commission's Scientific and Technical Guide-

---

16  The Commission on the Limits of the Continental Shelf was established pursuant to Annex II of UNCLOS. It consists of 21 experts in geology, geophysics, hydrography and geodesy who are elected by the state parties to the convention, with elections taking place every five years. More information on the commission can be obtained from its website: http://www.un.org/Depts/los/clcs_new/clcs_home.htm.

17  The term 'extended continental shelf' is in common use in respect of claims to continental shelf beyond 200 nautical miles. It is also the case that if Indonesia did not make a submission in respect of such areas, its maritime jurisdiction would in all probability not be 'extended' to include such continental shelf areas beyond 200 nautical miles from its baselines.

lines,[18] which were adopted on 13 May 1999. Article 76 of UNCLOS and the Scientific and Technical Guidelines are therefore the primary references for a coastal state wishing to delineate the outer limits of its extended continental shelf. On this basis, key components of a submission will include baselines, determination of the foot of the continental slope, the morphology or shape of the continental margin, the thickness of the sedimentary rocks comprising part of the continental margin as well as the bathymetry of the offshore areas in question.

Article 76 provides for two methods, or entitlement formulas, by which a coastal state may establish the existence of a continental margin forming part of its 'natural prolongation', beyond 200 nautical miles from its baselines. Both are measured from the foot of the continental slope. Establishing the location of the foot of the continental slope is therefore crucial to the application of these formulas. The foot of the continental slope is defined as the point of maximum change in gradient at the base of the continental slope.[19] According to the first of the two entitlement formulas, the limit of a coastal state's continental margin can be defined by reference to the thickness or depth of the sedimentary rocks overlying the continental crust. Specifically, this entitlement formula consists of a line delineating points where the thickness of the sediment is 1 per cent of the distance from the foot of the continental slope (known as the Gardiner Line).[20] The second method of establishing entitlement is through a distance formula consisting of a line 60 nautical miles distant from the foot of the continental slope (called the Hedberg Line) (article 76(4)). A coastal state may employ whichever of the two formulas is most advantageous to it, meaning that its outer limit may be composed of a combination of the two types of entitlement lines.

Once it has been established that the continental shelf does indeed extend beyond the 200-nautical-mile limit, two maximum constraints (or 'cut-off' lines) on the coastal state's outer continental shelf limits are applied. One of these cut-off lines is based on distance — 350 nautical miles from the coastal state's relevant baselines. The other is based on depth and distance — the 2,500-metre isobath plus 100 nautical miles (article 76(5)).[21] In a similar fashion to the application of the two entitle-

---

18   The guidelines set out the technical and scientific procedures to define the outer limits of an extended continental shelf. They are published in document CLCS/11, available at http://www.un.org/Depts/los/clcs_new/commission_documents.htm#Guidelines.

19   This is the case unless 'evidence to the contrary' exists (article 76(4b)).

20   The Gardiner Line was named after its principle architect, P.R.R. Gardiner (see Gardiner 1978).

21   The term 'isobath' refers to a depth contour line. Thus, the 2,500-metre isobath is the line connecting points with a depth of 2,500 metres.

*Figure 5.2    The outer limits of the continental shelf*

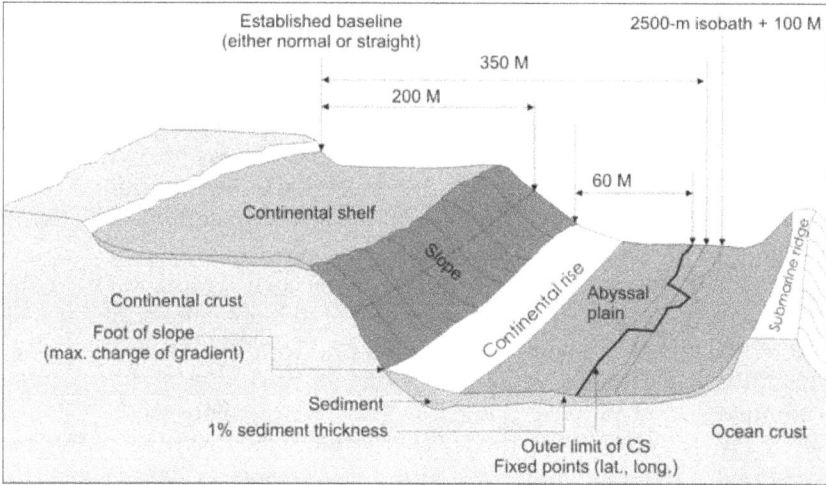

*Source:* Authors' research.

ment formulas, the coastal state may select the most advantageous option resulting from the combination of the two constraint lines. In practical terms, therefore, if the margin is broad and the coastal state can readily establish that its continental shelf extends well beyond the 200-nautical-mile limit, it may choose to apply whichever of the two cut-off lines is located further seaward and thus is most advantageous to the state in question. Article 76(7) also provides that the coastal state shall define the outer limits of its continental shelf where it extends beyond 200 nautical miles from its baselines 'by straight lines not exceeding 60 nautical miles in length, connecting fixed points, defined by coordinates of latitude and longitude'.

Delineation of the outer limits of the extended continental shelf, employing the above formulas and constraints, is illustrated in a three-dimensional perspective in Figure 5.2, which shows the idealized morphology of the seabed together with baselines and 200-nautical-mile limit lines. It also illustrates the slope, the foot of the continental slope, the continental rise and the abyssal plain where the outer limit of the extended continental shelf may lie, as well as the thickness of sediment and the location of the 2,500-metre isobath plus 100 nautical miles.

In light of the above provisions and requirements, it is not surprising that the process and procedure for delineating the outer limits of the extended continental shelf is challenging. As noted earlier, detailed geoscientific information is required and this requires expensive and time-consuming surveys, notably both bathymetric and seismic surveys. The objective of bathymetric surveys is to measure the depth of the ocean in such a way as to reveal the morphology of the ocean floor. The aim of seismic surveys, in contrast, is to define the thickness of sedimentary rock on the ocean floor (UNCLOS, article 76(4–6); CLCS/11 (see footnote 18); Arsana 2007). The information thus collected then needs to be analysed and interpreted, applied to fulfil the requirements of article 76 and presented in the form of a submission to the Commission on the Limits of the Continental Shelf. Clearly this process requires considerable scientific, technical, human and financial resources, including expertise in geology, hydrography, geophysics and geodesy.[22]

UNCLOS originally defined the deadline for submission by coastal states as 10 years after the entry into force of the convention for that state (Annex II, article 4). In retrospect it became clear that this deadline was unrealistically short, especially in view of the intricacies of the extended continental shelf process. What was perhaps originally envisaged as a relatively straightforward procedure has ultimately become one that is complex, challenging to apply and rigorously assessed by the commission. It is also the case that many more coastal states potentially have extended continental shelf rights than originally anticipated. Many of them are also developing states with many more pressing demands on the national purse, making the 10-year timeframe for submission even more testing.

The convention itself entered into force only on 16 November 1994, 12 years after its adoption in Montego Bay.[23] Accordingly, the first deadline for extended continental shelf submissions was set at 16 November 2004. However, in light of the relatively slow progress towards the preparation of submissions on the part of a number of interested coastal states, the deadline was pushed back, and the adoption of the commission's Scientific and Technical Guidelines on 13 May 1999 was instead taken as the commencement date of the 10-year 'clock'. As a result, the deadline for states that had become parties to UNCLOS before 13 May 1999 was shifted to 13 May 2009.[24] This was the deadline that was applicable to Indonesia, since it ratified the convention on 3 February 1986.

---

22  Arsana (2007) observes that a detailed survey conducted in one particular location, such as the area to the north of Papua, may cost up to US$1.4 million. This is comparable to the tuition fees for approximately 600 students studying geodetic engineering at Gadjah Mada University in Indonesia with an average five-year duration of study.

23  This was despite the fact that several states ratified it reasonably quickly. The convention came into force one year after the 60th instrument of its ratification was received by the secretary-general of the United Nations.

24  In 2001, a meeting of the state parties to UNCLOS decided that the 10-year time period referred to in article 4 of Annex II to UNCLOS 'shall be taken

Despite the deadline being pushed back, submissions were still relatively slow in being delivered to the commission. Indeed, by May 2008, one year before the new deadline set for many coastal states, only 11 submissions had been received by the commission, including just one from a developing state (Brazil). In recognition of this, and with the May 2009 deadline less than a year away, a meeting of state parties to UNCLOS in June 2008 made a further amendment. Although the May 2009 cut-off date was retained, the requirements to meet that deadline and 'stop the clock' were substantially eased. Thus, instead of submitting a complete submission (including comprehensive data and documents), coastal states would be allowed to submit 'preliminary information indicative of the outer limits of the continental shelf beyond 200 nautical miles and a description of the status of preparation and intended date of making a submission'.[25] Supporting data and other requirements could be provided at a later date.

The Russian Federation was the first coastal state to make a submission on the outer limits of its extended continental shelf. At the time of writing, the commission had made some recommendations regarding the Russian submission. It did, however, request additional information from Russia and that country is currently preparing for a resubmission.[26] The commission has provided recommendations on the submissions of Brazil covering four different regions. In April 2008, it provided recommendations that are set to lead to approximately 2.56 million square kilometres of continental shelf seaward of Australia's 200-nautical-mile limits being confirmed as Australian extended continental shelf (Geoscience Australia 2008). Similarly, Ireland has received recommendations confirming around 39,000 square kilometres of Irish seabed area. Recommendations delivered in August 2008 on New Zealand's submission will give that country approximately 1.7 million square kilometres of extended continental shelf (New Zealand Ministry of Foreign Affairs and Trade 2008). In March 2009, the commission delivered recommendations regarding the joint submission by France, Ireland, Spain and the United

---

to have commenced on 13 May 1999' considering that 'it was only after the adoption by the Commission of its Scientific and Technical Guidelines on 13 May 1999 that States had before them the basic documents concerning submissions in accordance with article 76, paragraph 8, of the Convention'. See SPLOS/72, available at http://www.un.org/Depts/los/meeting_states_parties/SPLOS_documents.htm.

25  See SPLOS/183, available at http://www.un.org/Depts/los/meeting_states_parties/SPLOS_documents.htm.

26  In a presentation to the UNCLOS Symposium in Oslo (7–8 August 2008), Victor Poselov stated that Russia might need three years to resubmit the outer limits of its extended continental shelf in the Arctic region.

Kingdom in the area of the Celtic Rise and the Bay of Biscay; the submission by Norway in the northeast Atlantic and the Arctic; and the submission by Mexico in respect of the western 'gap' in the Gulf of Mexico. It is testament to the importance states attach to the deadline that as this chapter was being finalized in May 2009, the number of submissions had ballooned to 50, with a further 41 sets of preliminary information also being delivered to the commission (see Table 5.1).[27]

## EXTENDING INDONESIA? THE SEARCH FOR THE OUTER LIMITS OF INDONESIA'S EXTENDED CONTINENTAL SHELF

Indonesia started assessing the potential for an extended continental shelf submission in 2001. Scientists in the National Coordinating Agency for Surveys and Mapping (Badan Koordinasi Survey dan Pemetaan Nasional, or Bakosurtanal) conducted a desktop study in 2003 which suggested that Indonesia could potentially make submissions relating to an extended continental shelf west of northern Sumatra, north of Papua and south of Sumba Island (Sutisna, Tripatmasari and Khafid 2005: 1, 5). The study used data already available from the Digital Marine Resource Mapping Project for 1996–99, coupled with data in the public domain, such as global sea-bottom topographic data from ETOPO2, publicly available worldwide bathymetric and topographic data, and sediment thickness data from the American National Geophysical Data Centre/ National Oceanic and Atmospheric Administration (Sutisna, Tripatmasari and Khafid 2005: 1).

For the area to the northwest of Sumatra, field surveys were conducted from 21 January to 25 February 2006 using the Indonesian *Baruna Jaya VIII* and German *Sonne–BGR* survey vessels (Bakosurtanal 2008a). Further field surveys were conducted from 3 to 21 October 2006 and from 11 October to 11 November 2006 in the area to the south of Nusa Tenggara. The first of these was a bathymetric survey using the *Baruna Jaya VIII* to determine the foot of the continental slope. The second was a seismic survey using the *Sonne–BGR* to obtain data to verify the sedimentary rock thickness on the seabed area to the south of Nusa Tenggara. These studies confirmed earlier global assessments by Prescott (1998), the US Geological Survey (n.d.) and the Commission on the Limits of the Conti-

---

27  For details of submissions and preliminary information, see http://www. un.org/Depts/los/clcs_new/clcs_home.htm. Information on the status of submissions is available on the same website; see the statements by the chair on progress in the work of the commission and other relevant press releases.

*Table 5.1    List of extended continental shelf submissions, 13 May 2009*[a]

| No. | Submitting state | Date of submission | Status |
|-----|------------------|--------------------|--------|
| 1 | Russian Federation | 20 December 2001 | RP |
| 2 | Brazil | 17 May 2004 | RP |
| 3 | Australia | 15 November 2004 | RP |
| 4 | Ireland | 25 May 2005 | RP |
| 5 | New Zealand | 19 April 2006 | RP |
| 6 | Joint submission by France, Ireland, Spain and United Kingdom | 19 May 2006 | RP |
| 7 | Norway | 27 November 2006 | RP |
| 8 | France | 22 May 2007 | BC |
| 9 | Mexico | 13 December 2007 | RP |
| 10 | Barbados | 8 May 2008 | BC |
| 11 | United Kingdom | 9 May 2008 | BC |
| 12 | Indonesia | 16 June 2008 | BC |
| 13 | Japan | 12 November 2008 | BC |
| 14 | Joint submission by Mauritius and Seychelles | 1 December 2008 | BC |
| 15 | Suriname | 5 December 2008 | NC |
| 16 | Myanmar | 16 December 2008 | NC |
| 17 | France | 5 February 2009 | NC |
| 18 | Yemen | 20 March 2009 | NC |
| 19 | United Kingdom | 31 March 2009 | NC |
| 20 | Ireland | 31 March 2009 | NC |
| 21 | Uruguay | 7 April 2009 | NC |
| 22 | Philippines | 8 April 2009 | NC |
| 23 | Cook Islands | 16 April 2009 | NC |
| 24 | Fiji | 20 April 2009 | NC |
| 25 | Argentina | 21 April 2009 | NC |
| 26 | Ghana | 28 April 2009 | NC |
| 27 | Iceland | 29 April 2009 | NC |
| 28 | Denmark | 29 April 2009 | NC |
| 29 | Pakistan | 30 April 2009 | NC |
| 30 | Norway | 4 May 2009 | NC |
| 31 | South Africa | 5 May 2009 | NC |
| 32 | Joint submission by Federated States of Micronesia, Papua New Guinea and Solomon Islands | 5 May 2009 | NC |
| 33 | Joint submission by Malaysia and Vietnam | 6 May 2009 | NC |

*continued*

*Table 5.1 (continued)*

| No. | Submitting state | Date of submission | Status |
|-----|------------------|--------------------|--------|
| 34  | Joint submission by France and South Africa | 6 May 2009 | NC |
| 35  | Kenya | 6 May 2009 | NC |
| 36  | Mauritius | 6 May 2009 | NC |
| 37  | Vietnam | 7 May 2009 | NC |
| 38  | Nigeria | 7 May 2009 | NC |
| 39  | Seychelles | 7 May 2009 | NC |
| 40  | France | 8 May 2009 | NC |
| 41  | Palau | 8 May 2009 | NC |
| 42  | Côte d'Ivoire | 8 May 2009 | NC |
| 43  | Sri Lanka | 8 May 2009 | NC |
| 44  | Portugal | 11 May 2009 | NC |
| 45  | United Kingdom | 11 May 2009 | NC |
| 46  | Tonga | 11 May 2009 | NC |
| 47  | Spain | 11 May 2009 | NC |
| 48  | India | 11 May 2009 | NC |
| 49  | Trinidad and Tobago | 12 May 2009 | NC |
| 50  | Namibia | 12 May 2009 | NC |

RP = recommendation provided; BC = being considered, NC= not yet considered.
a   Some states appear more than once in the table because they have made more than one submission. Norway, for example, has made two submissions, one relating to an area in the northeast Atlantic and Arctic, and the other to Bouvetøya and Dronning Maud Land in the Antarctic. France has made submissions covering areas in three regions: French Guiana and New Caledonia; the French Antilles and Kerguelen Islands; and La Réunion Island and Saint-Paul and Amsterdam islands. The United Kingdom has also made submissions covering areas in three regions: Ascension Island; the Hatton–Rockall Area; and the Falkland Islands, and South Georgia and the South Sandwich Islands. Ireland has made two submissions, one relating to the Porcupine Abyssal Plain and the other to the Hatton–Rockall Area.
*Source:* http://www.un.org/depts/los/clcs_new/commission_submissions.htm.

nental Shelf (Symonds 2008),[28] which all identified similar potential locations for the Indonesian extended continental shelf.

For the field surveys, the Indonesian team collaborated with agencies in Germany and Norway, which provided either survey equipment or human resource training (Arsana 2007). Financial support for the

---

28   Professor Symonds is Senior Adviser, Law of the Sea, at Geoscience Australia. He is also a member of the Commission on the Limits of the Continental Shelf, having been elected for the period 2002–07 and re-elected for the term 2007–12.

extended continental shelf project has been limited.[29] Proper field surveys require significant financial backing, but this proved difficult to secure in competition with other economic and development priorities, forcing the team to work with other states conducting research in Indonesia's maritime zones. For the area to the west of Sumatra, for example, data used in the submission drew on surveying conducted in the aftermath of the December 2004 tsunami for the purpose of disaster mitigation and management.[30] The data acquired in such ways were not always adequate, in terms of spatial resolution, for example. There were also, inevitably, political dimensions to the issue, for example in convincing members of the House of Representatives (Dewan Perwakilan Rakyat, or DPR) of the relative importance of the extended continental shelf to Indonesia's national interest and priorities.

Notwithstanding these challenges, the team managed to prepare a submission, which was presented to the commission on 16 June 2008.[31] It covered only the area to the northwest of Sumatra. However, Indonesia is currently preparing two further submissions regarding an area to the south of Nusa Tenggara and an area to the north of Papua (Bakosurtanal 2008b).

Figure 5.3 illustrates the area to the northwest of Sumatra, with particular attention to the proposed extended continental shelf. It shows that Indonesia's partial submission relates to a relatively small area of some 3,915 square kilometres — an area approximately equal in size to Madura island, located to the northeast of Java (Bakosurtanal 2008a). This is a modest proposal when compared with the extended continental shelf areas covered by other submissions, such as that of Australia (Arsana and Putro 2006; Geoscience Australia 2008; Schofield 2008: 9), suggesting that Indonesia has adopted a reasonably conservative approach. However, the Australian and Indonesian submissions cannot be compared directly, because the geographical and geological settings of the two countries are completely different. Most of Indonesia's northern maritime border abuts the maritime claims of its neighbours, while to the south the edge of the continental shelf is marked by the long, deep Java Trench, which runs relatively close to the Indonesian coast. Australia, in contrast, is bordered by only four neighbours (Indonesia, Papua New

---

29  Personal communication, Dr Khafid, Bakosurtanal, 18 August 2007. See also Sutisna, Tripatmasari and Khafid (2005: 1).
30  Personal communication, Dr Khafid, Bakosurtanal, 18 August 2007.
31  An executive summary of the submission is available at http://www.un.org/depts/los/clcs_new/submissions_files/submission_idn.htm. Consisting of 10 pages and two figures, it briefly highlights legal and technical aspects of the submission. It also contains an appendix listing the coordinates of the outer limits of the continental shelf.

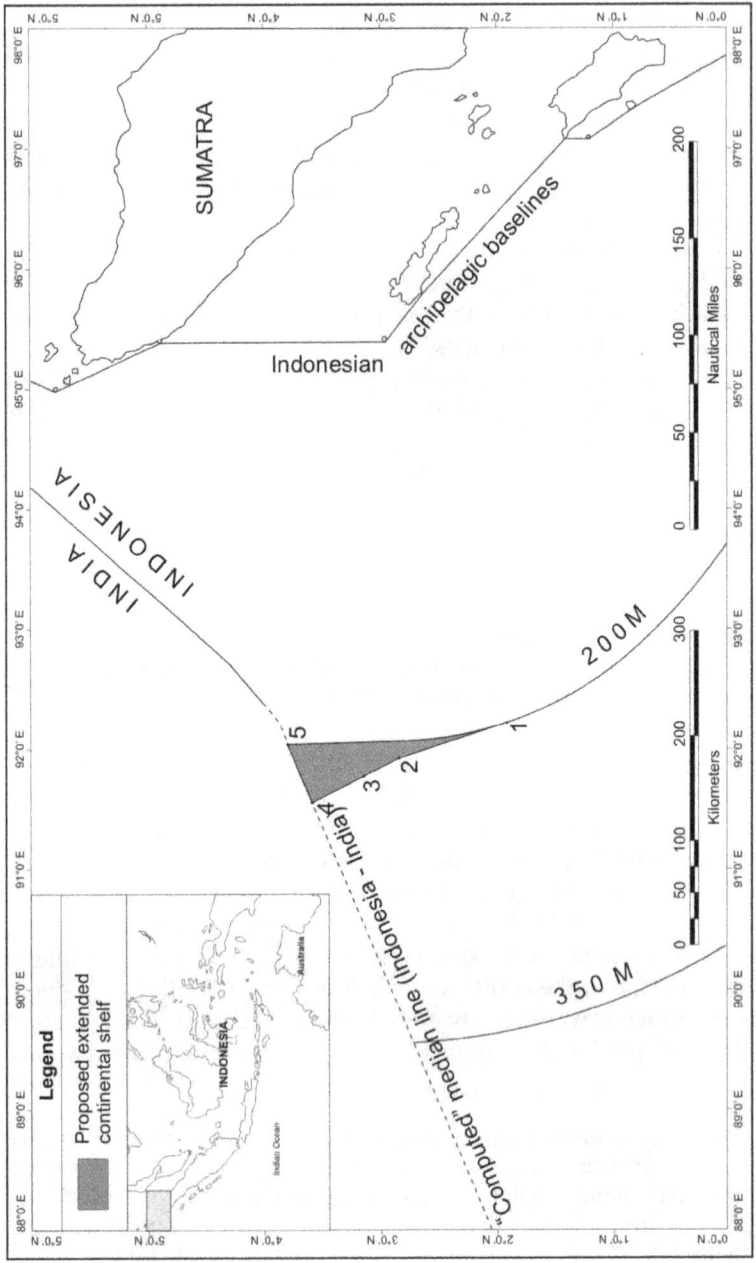

Figure 5.3  Proposed extended continental shelf to the northwest of Sumatra

Source: Based on Figure 2 in Republic of Indonesia (2008: 7).

86

Guinea, France and New Zealand) with whom it may exercise overlapping claims, and, crucially, as a continent it is naturally fringed with a broad continental shelf. This geographical position has helped Australia confirm its sovereign rights over an extremely large extended continental shelf entitlement.

Considering the fact that the preparation of the submission required considerable expenditure over a period of 10 years, the relatively small area proposed raises the question of whether the effort is worth the costs incurred. In addition, even if the commission were in due course to provide recommendations supporting Indonesia's submission, Indonesia would not possess full sovereignty over the additional continental shelf area beyond its 200-nautical-mile limit, but merely limited sovereign rights in respect of the seabed resources in the area. Nonetheless, while direct benefits are difficult to see in the short term, the technically complex process of preparing a submission represents an important investment for the future, particularly as knowledge and understanding of the potential utility of the seabed and subsoil is by no means complete and the area may yield as yet untapped and indeed unlooked-for resources in the future. Also, if Indonesia had not made a submission, it would in all probability have lost the opportunity to confirm its sovereign rights over the area of extended continental shelf under consideration. Once Indonesia confirms the outer limits of its extended continental shelf on the basis of the commission's recommendations, these outer limits will be final and binding. Fundamentally, therefore, the issue of the extended continental shelf is one that is linked to national sovereignty or, more precisely in respect of the continental shelf, sovereign rights. This factor in itself provides a powerful rationale for Indonesia to bear the burden of preparing a submission. In addition, the nation can in all likelihood expect to gain more area from the other two submissions that are still pending.[32]

Another important issue concerns the potential for Indonesia's submission to stir up or highlight disputes with neighbouring states. Indonesia shares maritime borders with 10 countries — India, Thailand, Malaysia, Singapore, Vietnam, the Philippines, Palau, Papua New Guinea, Australia and Timor-Leste — leading to overlapping maritime jurisdictional entitlements. Distances of less than 400 nautical miles from most of these neighbours significantly lessen Indonesia's extended continental shelf potential. For example, the submission for the area to the north-

---

32 Personal communication, Arif Havas Oegroseno, Director of Treaties for Political, Security and Territorial Affairs, Indonesian Ministry of Foreign Affairs, 20 September 2008.

west of Sumatra brought Indonesia's maritime boundary with India into play. Although Indonesia has confirmed to the commission that the area claimed is not subject to dispute with any other state (Republic of Indonesia 2008: 6), one of the points Indonesia is proposing is a fixed point on a 'computed median line' between Indonesia and India (see Figure 5.3).

Reference to this point could have proved problematic, since article 76 of UNCLOS does not mention a computed median line as one of the criteria for defining the outer limits of a continental shelf. In addition, point 4 is not part of an agreed line, but instead appears to have been calculated by Indonesia unilaterally.[33] Consequently, the potential existed for India to raise an objection (or at least to reserve its rights) and thus for a dispute to emerge. At the time of writing (over six months after the submission was made) no response had been provided by any state, including India. This suggests that India has no objection to Indonesia's use of its apparently unilaterally drawn computed median line. The commission's considerations will therefore be based solely on its interpretation and assessment of the submission.[34] Had India asked the commission not to consider the submission on the grounds that it might be prejudicial to future maritime boundary delimitation, it is quite possible that the commission would have acceded to that request.[35] For its own part, India made a partial submission to the commission on 11 May 2009.[36] However, the partial submission concerns only the Eastern Offshore Region (Bay of Bengal sector and western Andaman sector) and Western Offshore Region, and has nothing to do with India–Indonesia maritime boundaries.

---

33  Members of the Indonesian team have confirmed that Indonesia did not engage in any diplomatic communication with India concerning the submission (personal communication, Ms Tripatmasari, member of the Indonesian submission team from Bakosurtanal, 4 August 2008; personal communication, Arif Havas Oegroseno, 20 September 2008).

34  See also Rules of Procedure of the Commission on the Limits of the Continental Shelf (CLCS/40/Rev1), Annex III, available at http://www.un.org/Depts/los/clcs_new/commission_documents.htm#Guidelines.

35  Article 76(10) of UNCLOS contains an explicit guarantee that: 'The provisions of this article are without prejudice to the question of delimitation of the continental shelf between States with opposite or adjacent coasts'. However, where parts of a continental margin beyond 200 nautical miles from the relevant baselines are shared by more than one coastal state, one option open to the coastal states concerned is to consult and then indicate to the commission that they have no objection to it considering the others' submissions, without prejudice to the delimitation of a maritime boundary. Alternatively, a joint submission could be considered.

36  An executive summary of the submission is available at http://www.un.org/Depts/los/clcs_new/submissions_files/submission_ind_48_2009.htm.

## HOPES AND FUTURE CHALLENGES

At the time of writing, Indonesia was still at the beginning of a long journey to secure rights over its extended continental shelf. The commission will not start assessing its submission until at least its 23rd session, scheduled for 2009. It is therefore too early to judge how the commission will respond to the submission. Should the proposed area be confirmed as part of Indonesia's continental shelf, Indonesia will have an opportunity to explore the area and exploit its natural resources. Little is known about seabed resources in the area but they may be considerable.[37] Oil and gas are perhaps the best known and most economically attractive of such resources, but traditional seabed hydrocarbons are not the only potential resources of extended continental shelf areas. Other types of hydrocarbons (such as gas hydrates) as well as a range of other mineral resources may exist there, notably placer deposits of precious metals, polymetallic sulphides and manganese nodules.

It is also the case that coastal state rights over the continental shelf include rights over sedentary species (UNCLOS, article 77(4))[38] and this offers potential for bioprospecting. This emerging field describes activities related to the discovery and utilization of biological seabed resources for new applications. The extended continental shelf, like the deep seabed, may host unique environments inhabited by organisms with special characteristics (Mossop 2007). The behaviour and life processes of these organisms can be the subject of useful research with considerable value. Elsewhere, diverse environments have been discovered, such as hydrothermal vents, methane seeps and deep-sea sediments. These provide often extreme environments that lead to the evolution of unique deep-sea biological communities that possess properties with great potential to be explored and studied further for the development of science and technology.[39] In order to realize these opportunities, however, an intensive and sustained research effort is likely to be required on the part of the Indonesian government.[40] The commitment of the government in investing

---

37  See Allinson (2004), Herzig (2004), Kelly (2004) and Parson (2004).

38  Sedentary species include various types of molluscs (abalone, oysters, scallops), crustaceans (lobster, crab) and echinoderms (sea urchins, bêche-de-mer). On the UNCLOS definition of sedentary species, see footnote 41.

39  *Cryptotheca crypta* is an example of a unique organism from which *C-nucleosides* was isolated four decades ago. The latter provided the basis for clinical use of the first marine-derived anti-cancer agent, *cytarabine*. For a discussion of this, refer to Schwartsmann et al. (2001).

40  It can do this by supporting education generally and by funding relevant research programs specifically. In his state of the nation address on 15 August 2008, President Susilo Bambang Yudhoyono confirmed that the government

in research will play a vital role in Indonesia gaining more knowledge about the resource potential of its extended continental shelf. Should Indonesia's assertions over an extended continental shelf be confirmed, the government will face challenges concerning the management of these 'additional' areas as well as its extensive maritime jurisdiction more generally. As the extended continental shelf relates solely to the seabed and subsoil, a coastal state has sovereign rights over the seabed only, not the water column overlying it or the airspace above it. As a result, the seabed is subject to national jurisdiction while the water column falls within the high seas. This overlap of legal regimes may complicate the management of the extended continental shelf. For example, while Indonesia has exclusive access to utilize and manage natural resources within or attached to the seabed,[41] it has shared access, with other states, to resources in the water column (the high seas).[42] When a coastal state exploits the natural resources of its extended continental shelf, it is obliged to make annual payments to the International Seabed Authority (UNCLOS, article 82(1)),[43] which are then distributed to the states that are party to the convention (article 82(4)). As a result, revenue derived from the exploitation of the extended continental shelf's natural resources does not flow exclusively to the coastal state.

An expansion of Indonesia's maritime jurisdiction would also result in greater responsibilities. Indonesia's maritime surveillance system is relatively undeveloped, especially when compared with the systems implemented by developed neighbours such as Australia (Schofield, Tsamenyi and Palma 2008), and the number of vessels operated by the navy is small compared with the size of the maritime area subject to Indonesian jurisdiction. In fact, one of the biggest challenges for the government is to strengthen the capacity of its navy and other relevant agencies to secure Indonesia's large maritime area. Securing its maritime spaces even within a distance of 200 nautical miles has not been an easy task.

---

was proposing to raise the education budget to meet the 20 per cent requirement in the constitution. See the full text of his speech at http://www.presidensby.info/index.php/pidato/2008/08/15/971.html, accessed 13 September 2008.

41   UNCLOS, article 77(4) states that: 'The natural resources referred to in this Part consist of the mineral and other non-living resources of the seabed and subsoil together with living organisms belonging to sedentary species, that is to say, organisms which, at the harvestable stage, either are immobile on or under the seabed or are unable to move except in constant physical contact with the seabed or the subsoil'.

42   All states enjoy freedom of the high seas as set out in article 87 of UNCLOS.

43   The International Seabed Authority was provided for in UNCLOS but was not actually established until 1994. It is governed by Part XI, section 4 of the convention. For further details, see http://www.isa.org.jm/en/home.

Illegal, unreported and unregulated fishing, for example, remains a serious problem in several areas, partly because of the lack of regular patrols and the inadequacy of Indonesia's marine surveillance system relative to its enormous maritime jurisdiction.[44] The addition of an extended continental shelf would place further pressure on the navy and other agencies. By their very nature, extended continental shelf areas are remote and peripheral, a factor that exacerbates the enforcement challenge. If unauthorized bioprospecting activities were to be undertaken by a vessel—an act that could be termed 'biopiracy'—it would be difficult for the Indonesian authorities to detect such an offence, let alone apprehend the vessel in question.

## CONCLUDING REMARKS

Pursuant to Annex II of UNCLOS, Indonesia has made a partial submission to the Commission on the Limits of the Continental Shelf regarding an extended continental shelf. The submission covers an area of 3,915 square kilometres to the northwest of Sumatra. Indonesia is now preparing two more submissions: one for an area to the south of Nusa Tenggara and another for an area to the north of Papua. Indeed, with respect to the latter area, it has been reported that consultations have taken place between Indonesia, Papua New Guinea and the Federated States of Micronesia concerning the Eauripik Rise, an area of outer continental shelf shared between them.[45] Both Papua New Guinea and the Federated States of Micronesia duly provided the commission with preliminary information on the outer continental shelf areas encompassing parts of the Eauripik Rise.[46]

It is clearly not possible to predict the commission's likely recommendations on the partial submission, although it can be observed that

---

44  Purwanto, Secretary of the Directorate General for Monitoring and Control, Ministry of Marine Affairs and Fisheries, has identified several types of losses from illegal, unreported and unregulated fishing, including the loss of gas subsidies and fishery tax, the loss of resources and working fields, and the loss of value added from fishery processing and vessel maintenance. He calculates the losses from gas subsidies, fishery tax and resources alone at approximately Rp 1–4 billion per ship annually. Purwanto points out that Indonesia has only 21 vessels available for patrolling, but would need at least 70 to do the job properly ('Law enforcement decreases illegal fishing cases in country', *Jakarta Post*, 14 April 2008).

45  'PNG to extend border in deal', *Papua New Guinea Post-Courier*, 6 March 2009; SPC (2009).

46  See http://www.un.org/Depts/los/clcs_new/commission_preliminary.htm.

Indonesia's claim appears conservative and largely unproblematic. For Indonesia, the decision to secure a larger maritime area offers both challenges and hopes for the future. The challenges come from the complexity of the process of preparing and delivering a submission — something that clearly requires considerable resources and commitment. A successful submission would also raise hopes for opportunities in the shape of additional natural resources that may be able to be exploited in the future. Ultimately, however, making a submission and having areas of continental shelf seaward of the 200-nautical-mile limit confirmed as subject to Indonesian sovereign rights is not the end of the process. Possession of areas of extended continental shelf will lead to potentially daunting challenges in terms of the future administration and management of these remote and subsurface 'extensions' to Indonesia.

## REFERENCES

Allinson, G.J. (2004), 'New developments in oil exploration in the northeast Atlantic', in M.H. Nordquist, J.H. More and T.H. Heidar (eds), *Legal and Scientific Aspects of Continental Shelf Limits*, Martinus Nijhoff Publishers, Leiden, pp. 421–2.

Arsana, I M.A. (2007), 'The delineation of Indonesia's outer limits of its extended continental shelf and preparation for its submission: status and problems', paper for the UN–Nippon Fellowship, New York.

Arsana, I M.A. and B.A.W. Putro (2006), 'The Australia's submission of the extended continental shelf (ECS): a study on its impact to the Indonesia–Australia maritime boundaries and Indonesian potential claim over ECS', paper delivered to the 31st Annual Meeting of the Indonesian Association of Geophysicists, Semarang, 13–15 November.

Bakosurtanal (Badan Koordinasi Survey dan Pemetaan Nasional) (2008a), 'Indonesia klaim wilayah seluas Pulau Madura' [Indonesia claims a new area the size of Madura Island], 15 February, available at http://www.bakosurtanal. go.id/?m=30&p=3&view=178, accessed 20 September 2008.

Bakosurtanal (Badan Koordinasi Survey dan Pemetaan Nasional) (2008b), 'Tim LKI serahkan dokumen submisi ke Deplu' [Indonesian continental shelf team hands submission document to the Ministry of Foreign Affairs], 7 May, available at http://www.bakosurtanal.go.id/?m=30&p=2&view=200, accessed 11 July 2008.

Gardiner, P.R.R. (1978), 'Reasons and methods for fixing the outer limit of the legal continental shelf beyond 200 nautical miles', *Iranian Review of International Relations*, 11–12: 145–70.

Geoscience Australia (2006), 'Maritime boundary definitions', available at http:// www.ga.gov.au/oceans/mc_amb-bndrs.jsp, accessed 22 April 2008.

Geoscience Australia (2008), 'UN confirms Australia's extended marine jurisdiction', April, Canberra, available at http://www.ga.gov.au/news/archive/ 2008/april/, accessed 22 April 2008.

Herzig, P.M., 'Seafloor massive sulfide deposits and hydrothermal systems', in M.H. Nordquist, J.H. More and T.H. Heidar (eds), *Legal and Scientific Aspects of Continental Shelf Limits*, Martinus Nijhoff Publishers, Leiden, pp. 431–56.

Kelly, P.L. (2004), 'Deepwater oil resources: the expanding frontier', in M.H. Nordquist, J.H. More and T.H. Heidar (eds), *Legal and Scientific Aspects of Continental Shelf Limits*, Martinus Nijhoff Publishers, Leiden, pp. 413–19.

Mossop, J. (2007), 'Protecting marine biodiversity on the continental shelf beyond 200 nautical miles', *Ocean Development and International Law*, 38(3): 283–304.

New Zealand Ministry of Foreign Affairs and Trade (2008), 'UN confirms NZ's extended seabed claim', available at http://www.mfat.govt.nz/Features/NZs-extended-seabed-claim.php, accessed 5 October 2008.

Parson, L. (2004), 'Non-hydrocarbon resources', in M.H. Nordquist, J.H. More and T.H. Heidar (eds), *Legal and Scientific Aspects of Continental Shelf Limits*, Martinus Nijhoff Publishers, Leiden, pp. 423–9.

Prescott, J.R.V. (1998), 'National rights to hydrocarbon resources of the continental margin beyond 200 nautical miles', in G.H. Blake, M.A. Pratt, C.H. Schofield and J. Allison Brown (eds), *Boundaries and Energy: Problems and Prospects*, Kluwer Law International, The Hague, pp. 51–82.

Republic of Indonesia (2008), *Continental Shelf Submission of Indonesia; Partial Submission in Respect of the Area of North West of Sumatra: Executive Summary*, available at http://www.un.org/depts/los/clcs_new/submissions_files/submission_idn.htm.

Schofield, C. (2008), 'Australia's final frontiers? Developments in the delimitation of Australia's international maritime boundaries', *Maritime Studies*, 158 (January–February): 2–21.

Schofield, C., M. Tsamenyi and M.A. Palma (2008), 'Securing maritime Australia: developments in maritime surveillance and security', *Ocean Development and International Law*, 39(1): 94–112.

Schwartsmann, G., A. Brondani da Rocha, R.G.S. Berlinck and J. Jimeno (2001), 'Marine organisms as a source of new anticancer agents', *Lancet Oncology*, 2(4): 221–5.

SPC (Secretariat of the Pacific Community) (2009), *Regional Maritime Information Bulletin*, 69.

Sutisna, S., Tripatmasari and Khafid (2005), 'Indonesian searching for it's continental shelf outer limits', paper presented to the ABLOS Tutorials and Conference, Marine Scientific Research and the Law of the Sea: The Balance between Coastal State and International Rights, 10–12 October, Monaco, available at http://www.gmat.unsw.edu.au/ablos/ABLOS05Folder/SutisnaPaper.pdf, accessed 15 September 2008.

Symonds, P.A. (2008), 'Extended continental shelf', lecture at the University of Wollongong, Wollongong, July.

Tsamenyi, M.B., C.H. Schofield and B. Milligan (2008), 'Navigation through archipelagos: current state practice', in M. Nordquist, T.B. Koh and J.N. Moore (eds), *Freedom of the Seas, Passage Rights and the 1982 Law of the Sea Convention*, Martinus Nijhoff, The Hague, pp. 413–54.

US Department of State (1971), *International Boundary Study, Series A, Limits in the Seas. Straight Baselines: Indonesia*, Office of the Geographer, Bureau of Intelligence and Research, Washington DC, available at http://www.state.gov/documents/organization/61544.pdf.

US Geological Survey (n.d.), 'USGS geological research activities with U.S. Minerals Management Service', available at http://geology.usgs.gov/connections/mms/landscapes/law_of_sea.htm, accessed 15 September 2008.

# 6  INDONESIAN PORT SECTOR REFORM AND THE 2008 SHIPPING LAW

*David Ray*

As the world's largest archipelagic nation, Indonesia requires a well-developed and efficiently run ports sector. This is because producer competitiveness in national and international markets, internal distribution efficiency and, more generally, national economic cohesiveness and integrity are influenced to a significant extent by port sector performance. Despite its obvious critical importance to the national economy, Indonesia does not have a port system that performs well from the perspective of its users. Indonesia's main port terminal, the Jakarta International Container Terminal (JICT), has been shown to be one of the least efficient of the main terminals in Southeast Asia in terms of productivity and unit costs (Ray 2003), and yet it is one of the better performing Indonesian ports. Performance indicators for all the major commercial ports suggest the entire port system is highly inefficient and in urgent need of upgrading. Berth occupancy rates, average turnaround times and working time as a percentage of turnaround time are well below international standards, suggesting that vessels are spending too much time at berth, or in queues outside ports.

Geographic factors such as the lack of deep-water harbours and the inland location of many ports on rivers that require near constant dredging are important constraints to port performance. But arguably the greatest constraint to development is the overall lack of private sector participation (investment) and competition in the ports system. This is in large part due to the dominance of the state in the provision of port services through the activities of four state-owned enterprises, the Indonesian Port Corporations or IPCs (Pelabuhan Indonesia), as well as deficiencies in the current legal and regulatory environment, which effectively constrains competition both within and between ports.

Some of these issues were addressed in Law No. 17/2008 on Shipping, which provides the foundation for comprehensive reform of the Indonesian port system. Most notably, the law removes the legislated state-sector monopoly on ports and opens the door for new participation by the private sector. This could lead to the injection of much needed competition, putting downward pressure on prices and driving general improvements in port services. Transforming the Indonesian ports system, however, will be a long and arduous process. The 2008 Shipping Law is a crucially important and positive first step, but much remains to be done with regard to developing supporting institutions, regulations and planning documents. Until this regulatory and institutional framework is in place, investors face a policy vacuum, unsure of what processes must be pursued, and what approvals and permits must be obtained from which agencies.

## THE CURRENT CONDITION OF INDONESIAN PORTS

Indonesia's ports are currently governed by Law No. 21/1992 on Shipping and its supporting regulations. The new regulatory regime, under the umbrella of the 2008 Shipping Law, will not be implemented fully until 2011. Under the current regime, ports are organized into a hierarchic system of approximately 1,700 ports. The 111 ports, including the 25 main 'strategic' ports, deemed to be commercial ports are controlled by the four state-owned IPCs (Pelindo, I, II, III and IV), whose geographic coverage is outlined in Table 6.1. In addition there are approximately 614 non-commercial ports that tend to be unprofitable and are of little strategic value, as well as approximately 1,000 'special purpose' or dedicated private ports that serve the needs of individual companies (both private and state owned) in a number of industries, including mining, oil and gas, fishing and forestry. Some of these ports have facilities that are appropriate only for one commodity or group of commodities (for example, chemicals) and have limited capacity to accommodate third-party cargo. Others, however, have facilities that are appropriate for a broad range of commodities, in some cases including containerized cargo.

Currently the IPCs enjoy a legislated monopoly over the main commercial ports as well as regulatory authority over private sector ports. In almost all of the main ports, the IPCs act as both sole operator and port authority, dominating the supply of all major port services. They are in charge of port waters (including dredged channels and basins); vessel traffic movement, anchoring and berthing; pilotage and towage (tugboats); port facilities for stevedoring and animal handling; warehouses and stacking yards; and container, bulk and passenger terminals. They

*Table 6.1   Main ports administered by the Indonesian Port Corporations*

| IPC | Provinces covered | Main ports administered |
|---|---|---|
| Pelindo I | Aceh, North Sumatra, Riau | Belawan, Pekanbaru, Dumai, Tanjung Pinang, Lhokseumawe |
| Pelindo II | West Sumatra, Jambi, South Sumatra, Bengkulu, Lampung, Jakarta | Tanjung Priok, Panjang, Palembang, Teluk Bayur, Pontianak, Cirebon, Jambi, Bengkulu, Banten, Sunda Kelapa, Pangkal Balam, Tanjung Pandan |
| Pelindo III | Central Kalimantan, South Kalimantan, West Nusa Tenggara, East Nusa Tenggara | Tanjung Perak, Tanjung Emas, Banjarmasin, Benoa, Tenau/ Kupang |
| Pelindo IV | Sulawesi, Maluku, Papua | Makassar, Balikpapan, Samarinda, Bitung, Ambon, Sorong, Biak, Jayapura |

*Source:* PDP Australia (2005: 5).

also manage associated services, such as the provision of electricity, fresh water, garbage disposal and telephone services for vessels; land space for offices and industrial estates; and port training and medical centres.

Current legislation prevents the private sector from competing directly with the incumbent IPCs and other elements of the governance structure ensure that there is no competition between the IPCs, which are required by law to subsidize each other to ensure overall financial sustainability and to meet their public service obligations (Patunru, Nurridzki and Rivayani 2007). Under this system, profitable ports are required to subsidize unprofitable ones, further reducing performance incentives. In addition, port tariffs, which are largely determined by the central government, are imposed in a standard manner across ports, further reducing opportunities for competition. This is particularly significant where two ports share a contestable hinterland, as in the case of the ports of Tanjung Emas in Semarang and Tanjung Perak in Surabaya, both of which are operated by Pelindo III.

### Port traffic

About 90 per cent of Indonesia's external trade is transported by sea, and almost of all of the non-bulk trade (such as containers) is transhipped through Singapore and, increasingly, the Malaysian port of Tanjung Pelepas. Indonesia does not have its own transhipment port

Figure 6.1    Total port traffic handled by Indonesian ports, 2002–06
(million tonnes)[a]

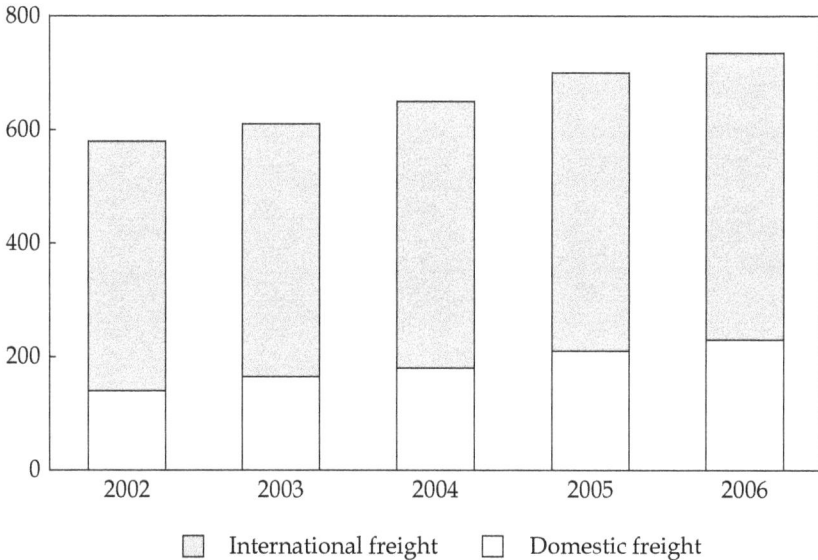

a   Data for 2006 are estimates.
*Source:* Ports and Dredging Directorate, Ministry of Transport.

capable of accommodating large transoceanic vessels, despite long-held government plans to develop such facilities in Bojonegara (just west of Jakarta), in Bitung (in North Sulawesi) and elsewhere in the archipelago. Even most of Indonesia's intra-Asia trade must be transhipped through regional hubs.

Ministry of Transport data show that the total tonnage handled by Indonesian ports grew from 582 million tonnes in 2002 to 736 million tonnes in 2006, or at an average annual rate of around 6 per cent (Figure 6.1). Over this period domestic freight grew at a rate of 11.5 per cent per annum, or nearly three times the rate for international freight of 4.1 per cent. Growth in domestic freight was particularly robust in eastern Indonesia during this period. In absolute terms, international and domestic freight each increased by around 77 million tonnes during the four years to 2006.

In the 11 main container terminals (that is, those provided with container cranes and declared 'container terminals' by the Ministry of Transport), container volumes increased by 1 million twenty-foot equivalent units (TEUs)—a TEU being the equivalent of a 20-foot container—between 2005 and 2007, representing an average annual growth rate of around 12 per cent (Table 6.2). The port of Tanjung Priok in Jakarta

*Table 6.2    Container volumes handled by the 11 main IPC ports, 2005–07 (no.)*

| Container port | 2005 | 2006 | 2007 |
|---|---|---|---|
| **Belawan (Medan)** | | | |
| Containers | 217,629 | 237,703 | 251,144 |
| TEUs | 281,106 | 304,002 | 320,515 |
| **Palembang** | | | |
| Containers | 60,805 | 65,648 | 76,893 |
| TEUs | 65,879 | 70,338 | 82,546 |
| **Panjang** | | | |
| Containers | 82,994 | 70,586 | 67,825 |
| TEUs | 93,164 | 81,545 | 79,767 |
| **Multi Terminal Indonesia (Jakarta)** | | | |
| Containers | 192,005 | 151,842 | 96,888 |
| TEUs | 295,477 | 222,762 | 135,019 |
| **JICT (Jakarta)** | | | |
| Containers | 994,352 | 1,085,977 | 1,212,564 |
| TEUs | 1,470,467 | 1,619,495 | 1,821,292 |
| **Koja (Jakarta)** | | | |
| Containers | 382,004 | 391,582 | 478,907 |
| TEUs | 573,410 | 583,065 | 702,199 |
| **Pontianak** | | | |
| Containers | 125,033 | 129,375 | 131,619 |
| TEUs | 132,273 | 138,991 | 143,443 |
| **Tanjung Perak (Surabaya)** | | | |
| Containers | 762,143 | 743,445 | 799,966 |
| TEUs | 1,073,385 | 1,051,960 | 1,113,478 |
| **Tanjung Emas (Semarang)** | | | |
| Containers | 211,443 | 219,965 | 233,582 |
| TEUs | 353,675 | 370,108 | 385,095 |
| **Makassar** | | | |
| Containers | | | |
| TEUs | 238,394 | 255,998 | 302,043 |
| **Bitung** | | | |
| Containers | | | |
| TEUs | | 44,958 | 55,623 |
| **Total** | **4,061,161** | **4,698,264** | **5,085,397** |
| **Annual growth rate (%)** | | **15.7** | **8.2** |

*Source:* Ports and Dredging Directorate, Ministry of Transport.

accounted for approximately half of all container throughput. In 2007, total container volume for the four terminals in the port was just under 3 million TEUs.[1] A number of important international trends in global sea freight are likely to affect the performance of Indonesia's ports. The first of these is what Penfold (2007) refers to as the ongoing 'size-based revolution in container ships', describing the use of ever larger vessels to enjoy lower per unit transport costs. Recent data show that a vessel of 12,000 TEUs on the Europe–East Asia route would generate an 11 per cent cost saving per container compared with an 8,000 TEU vessel, and a 23 per cent saving compared with a 4,000 TEU vessel (ESCAP 2007). Increasingly, the main transoceanic routes are being dominated by large vessels of over 12,000 TEUs. The smaller vessels of 5,000–8,000 TEUs previously used on these main trunk routes are being displaced to regional feeder routes. In the Indonesian context, this means that the smaller regional ports (which would include the main commercial ports in Indonesia) will require deeper channel draft and basin depth, bigger and faster cranes and improved cargo handling so that they can accommodate larger vessels. In addition, the presence of larger vessels on regional feeder routes will put added pressure on local shipping companies to upgrade their fleet of relatively small and ageing ships.[2]

The second key trend concerns the robust growth in international seafreight traffic and the impact of this growth upon regional ports. Over the past two decades international merchandise trade has grown at about 1.5–2 times the rate of growth of the global economy. Due to the increasing rate of containerization of freight, container trade grew at over twice the average annual growth rate of other maritime trade over the same period (ESCAP 2007). The most rapid growth in container volumes has occurred in East Asia, which now commands over half the world's container traffic. As noted by Kruk (2008) and others, the capacity of regional container terminals to handle this traffic is now reaching critical levels.[3]

---

1   This includes data from the Mustika Alam Lestari (MAL) container terminal in Jakarta (not shown in Table 6.2), which handled approximately 300,000 TEUs in 2007.

2   Of the 36 container vessels registered in Indonesia in 2005, 34 had a capacity of less than 1,500 TEUs and more than half were over 20 years in age (PDP Australia 2005).

3   Kruk (2008) refers to data in Drewry's 2005 *Annual Review of Global Container Terminal Operations*, which calculates regional capacity utilization based on (1) confirmed plans for expansion and (2) unconfirmed plans for expansion. Reflecting the overcapacity of regional container facilities, the figures for Northeast Asia were 109 per cent for confirmed expansions and 105 per cent for unconfirmed expansions, while the figures for Southeast Asia were 108 per cent and 91 per cent respectively.

The third important trend is the increasing role of the private sector in developing and operating container terminals. This is especially the case in developing countries where the public sector can no longer finance investment in new and expanded capacity (World Bank 2001). Since the early 1990s, the private sector has invested nearly US$33 billion in developing-country seaports, 44 per cent of them in the East Asia–Pacific region. With cargo volume increases outpacing the growth in terminal capacity, investors view the port sector as an attractive option, and foreign investors are paying 2–3 times as much for ports as they were in the late 1990s (Kruk 2008). Notwithstanding some partial, and some would say poorly managed, privatizations in the late 1990s and early 2000s, recent flows of international investment into seaports have largely bypassed Indonesia.

### Indicators of port performance

The most recent data available comparing Indonesia's performance with that of other countries in the region are from 2002 and are limited to the main trade gateway, Jakarta (Figure 6.2). While the data are a few years out of date, they nevertheless illustrate the relative lack of competitiveness of Indonesia's main port. Interviews with various international shippers indicate that this lack of competitiveness remains: Jakarta is both expensive and inefficient.

Time delays in the major port are now a matter of great concern to the main shipping lines. In 2002, Jakarta was achieving roughly 30–40 container moves per hour. Technical and operational improvements saw productivity increase to approximately 60 moves per hour by mid-2007. However, increased container traffic and port congestion coupled with problems associated with labour issues and customs delays saw productivity fall to around 40–45 moves per hour in early to mid-2008.

This is less than half the productivity rate in the Singapore transhipment port, which is currently working at around 100–110 moves per hour.[4] Due to delays in cargo handling, the main shipping lines report that vessels must often leave the port of Jakarta before they are fully loaded in order to keep to their published schedules. This means that

---

4  It is debatable whether it is appropriate to compare container terminals in Jakarta, which at most provide 2–3 cranes per ship, with those in Singapore or Tanjung Pelepas, where ships are serviced by 3–5 cranes. On a per crane basis, the main container terminals in Indonesia are achieving 18–22 moves per hour, while those in Singapore and Tanjung Pelepas are achieving at least 30–35 moves per hour.

Figure 6.2    Regional competitiveness of the port of Jakarta, 2002

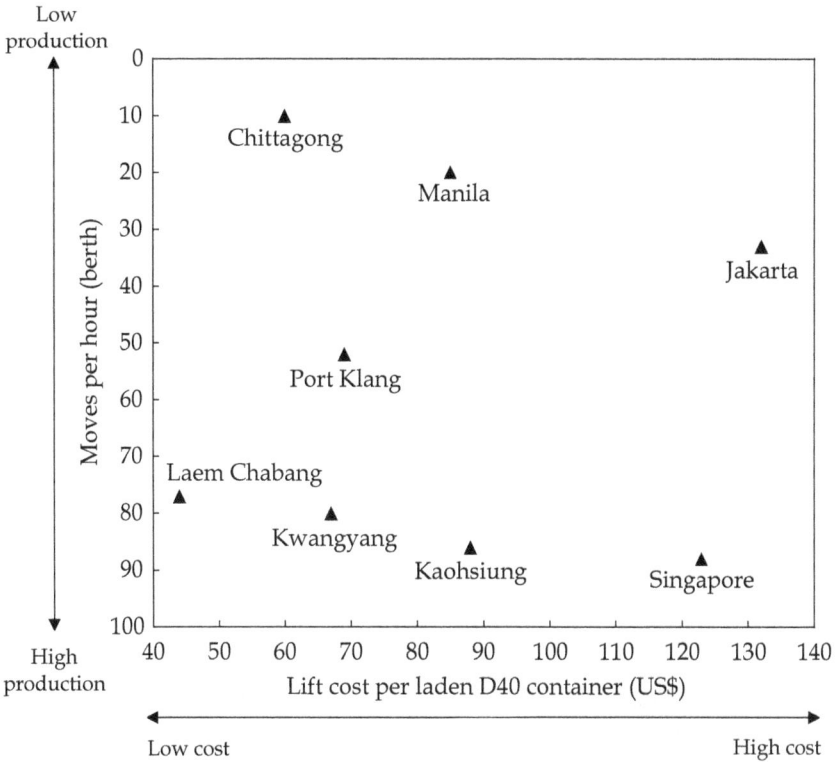

Source: Overseas Shippers Representatives Association (OSRA), quoted in Ray (2003).

the shippers incur various costs, inluding redressing charges,[5] the cost of procuring space on third-party feeder vessels and the cost of wasted space on their own feeder vessels.[6] As a result, these shipping lines are cutting back on capacity planned for the Jakarta port.[7]

Indonesia's international shippers enjoy very competitive (transhipment) port services in Singapore and Malaysia but must pay high feeder costs due mainly to the high port costs in Indonesia. Supply chain analy-

5  'Redressing' is the term used to describe a situation where containers planned for a particular vessel must be rescheduled for another vessel.
6  Feeder vessels operate between smaller subsidiary ports. They transfer their cargoes to large 'mother' vessels that operate on the main trunk routes.
7  This paragraph draws heavily on information obtained through interviews with country managers of major international shipping lines in Jakarta (April and May 2008).

sis shows that the cost of accessing regional hubs represents a dispro-
portionate amount of total international freight costs. Carana Corporation (2004) estimates that 20–50 per cent of the international freight costs
for exports are typically incurred in the first 1,000 miles of accessing a
regional hub. In one example, the 600 miles from the port of Semarang
(Central Java) to Singapore represents 10 per cent of the distance, but
over 45 per cent of the total freight costs, to export furniture to the end
market in Valencia, Spain.

Obtaining performance data on Indonesia's main gateway port is difficult, but data for most of Indonesia's other strategic ports are available.
From the data for the 19 ports for which complete data are available,
it can be seen that the delivery of port services to users has been poor,
and that there has been little improvement since the late 1990s. This is
reflected in a number of key performance indicators, such as berth occupancy rate, vessel turnaround time and working time ratio (Table 6.3).

Overall, the simple average for the berth occupancy rate for these ports
was 57.6 per cent in 2006, down from 65.0 per cent in 1999 but nevertheless well beyond what Nathan Associates (2001) and others regard as the
maximum internationally acceptable standard of 40 per cent. This raises
concern that growth in container volumes, without sufficient upgrades
in capacity, will lead to increased delays and waiting times for vessels.

Average turnaround time (a measure aggregating all time required
at port, including waiting time, approach time, idle time and so on) also
suggests poor port performance, with vessels requiring an average of 82
hours (3.4 days) in port in 2006, up from 79 hours in 1999. For the complete list of 25 strategic ports (including Pelindo II ports), turnaround
time in 2006 for domestic shipping was 74 hours (3.1 days), falling to 65
hours (2.7 days) in 2007.

Effective working time as a proportion of turnaround time averaged
44.5 per cent in 2005/06, which means that for the time a vessel was in
port, it was only being serviced (that is, loaded or unloaded) for less
than half that time. The same figure for 1999 was slightly higher at 44.7
per cent, suggesting that there has been little or no improvement in this
important indicator in recent years.[8]

The simple conclusion to be drawn from the above analysis is that the
Indonesian cargo fleet is spending too much time sitting idle or waiting
in port. The average sail time between the 19 ports listed in Table 6.3 and

---

8  Some of the ports on this list, such as Samarinda, Palembang and Pekanbaru,
   are river ports with longer approach times. However, removing approach
   time (AT) from the turnaround time (TRT) denominator has little impact on
   working time ratios.

*Table 6.3   Performance data for 19 major ports: domestic cargo*[a]

| Port | 1999 | 2006 | 1999 | 2006 | | | | | | |
|---|---|---|---|---|---|---|---|---|---|---|
| | BOR (%) | BOR (%) | TRT (hrs) | TRT (hrs) | WT (hrs) | PT (hrs) | AT (hrs) | NOT (hrs) | ET (hrs) | IT (hrs) |
| **Pelindo I** | | | | | | | | | | |
| Belawan | 62.7 | 52.4 | 77.9 | 72.6 | 1.4 | 16.6 | 1.7 | 22.4 | 29.8 | 0.9 |
| Dumai | 73.6 | 74.0 | 83.4 | 81.5 | 4.2 | 26.8 | 9.6 | 11.4 | 27.3 | 2.4 |
| Lhokseumawe | 43.2 | 22.4 | 88.8 | 62.7 | 0.8 | 5.8 | 1.3 | 25.8 | 27.4 | 1.6 |
| Pekanbaru[b] | 59.2 | 51.3 | 109.9 | 96.5 | 1.4 | 14.5 | 11.4 | 45.4 | 22.5 | 1.2 |
| Tanjung Pinang | 82.9 | 90.3 | 84.4 | 82.9 | 0.0 | 2.3 | 2.0 | 58.4 | 16.0 | 4.2 |
| **Pelindo II** | | | | | | | | | | |
| Banten | 41.6 | 39.1 | 57.9 | 65.1 | 1.0 | 0.8 | 7.8 | 34.5 | 21.1 | 0.0 |
| Palembang[b] | 62.9 | 34.7 | 73.6 | 61.8 | 0.1 | 0.0 | 17.7 | 20.0 | 23.3 | 0.7 |
| **Pelindo III** | | | | | | | | | | |
| Banjarmasin | 81.0 | 74.7 | 55.0 | 52.0 | 1.0 | 1.0 | 6.0 | 23.0 | 21.0 | 0.0 |
| Benoa | 60.1 | 56.0 | 22.0 | 137.0 | 0.0 | 0.0 | 1.0 | 122.0 | 14.0 | 0.0 |
| Tenau/Kupang | 74.4 | 65.7 | 79.0 | 167.0 | 10.0 | 1.0 | 6.0 | 65.0 | 85.0 | 0.0 |
| Tanjung Emas | 79.0 | 27.8 | 51.0 | 77.0 | 1.0 | 2.0 | 2.0 | 11.0 | 49.0 | 12.0 |
| Tanjung Perak | 63.0 | 69.0 | 99.0 | 38.0 | 0.0 | 5.0 | 4.0 | 9.0 | 20.0 | 0.0 |
| **Pelindo IV** | | | | | | | | | | |
| Ambon | 60.2 | 54.2 | 62.1 | 54.8 | 0.1 | 0.3 | 0.3 | 24.0 | 29.6 | 0.6 |
| Biak | 71.2 | 49.5 | 96.0 | 80.0 | 1.0 | 0.0 | 1.0 | 10.0 | 67.0 | 1.0 |
| Bitung | 65.1 | 70.2 | 95.6 | 60.5 | | 0.6 | 0.4 | 28.0 | 31.6 | 0.0 |
| Jayapura | 65.2 | 70.9 | 164.5 | 103.5 | 0.4 | 0.1 | 0.5 | 23.7 | 33.9 | 44.6 |
| Makassar | 53.8 | 43.2 | 66.7 | 124.3 | 0.0 | 0.0 | 3.0 | 15.2 | 93.4 | 12.6 |
| Samarinda[b] | 64.0 | 68.9 | 93.0 | 88.8 | 7.3 | 0.0 | 5.0 | 10.0 | 59.2 | 7.3 |
| Sorong | 72.4 | 80.0 | 38.3 | 50.0 | 6.0 | 0.0 | 1.0 | 20.0 | 22.0 | 1.0 |
| **Average** | **65.0** | **57.6** | **78.8** | **81.9** | **2.0** | **4.0** | **4.3** | **30.4** | **36.5** | **4.7** |

BOR = berth occupancy rate; TRT = turnaround time; WT = waiting time; PT = postponement time (caused by port administration); AT = approach time; NOT = down time; ET = effective working time; IT = idle time; hrs = hours.

a   Note that data are unavailable for four of the Pelindo II ports, as well as Balikpapan and Batam.

b   Denotes a river port.

*Source:* Ministry of Transport (2006).

the main feeder ports of Jakarta and Surabaya is 1–2 days.[9] This information combined with the data on turnaround times suggests that many of Indonesia's domestic cargo vessels are spending at least half and perhaps as much as three-quarters of their time in port.

### Key factors contributing to poor port performance

Port depth appears to be a major problem in virtually every port in Indonesia. The country has very few natural deep-water harbours and a river system prone to serious siltation that restricts port depth. For many ports, continuous dredging is a very expensive reality. Where dredging is not feasible, as is the case with the river port of Samarinda, vessels often have to wait until high tide to enter the port, which leads to more non-working time for vessels. Physical geography is particularly constraining for the country's main ports on the northern coast of Java, which service the most populous and industrialized regions of the country. This is due to the highly alluvial and unstable coastal soils/seabeds coupled with shallow coastal waters. The port of Semarang, the main seaport for Central Java, is particularly problematic in this regard as it is sinking at a rate of 7–12 centimetres per year and for many days of the month large parts of the port are under water. Every 7–10 years the container terminal must go through an expensive and time-consuming exercise to elevate the main wharf and storage area.

The fact that almost all the major Indonesian ports are located close to large urban areas where access is via busy roadways represents another constraint. Congestion problems are often exacerbated by the arrival of passenger vessels, as only a few regional ports have separate facilities for cargo and passenger ships. In ports with high berth occupancy rates, the simultaneous presence of passenger and cargo vessels results in even further delays, and increases the total turnaround time for cargo vessels.

Port performance is also constrained by poor infrastructure. Many regional ports lack container facilities, requiring shipping lines to use their own gear, stored both on board and at port. Only 16 of the 111 commercial ports have container-handling equipment of some type. Recently there were long shipping delays at certain ports, most notably Panjang in Lampung and Belawan in North Sumatra, caused by the breakdown of key port-side equipment (such as gantry cranes) and delays in getting replacement parts.[10] Lack of space for container storage and stuffing

---

9   'Re-evaluate TPK Koja', *Indonesia Shipping Gazette*, 5 May 2008: 15.
10  In Panjang, for example, it was reported that damaged cranes were causing delays of up to a day and half in May 2008 ('Port inspection as culprit in smuggling', *Indonesia Shipping Times*, July 2008: 14).

is another problem confronting most Indonesian ports. This often man-dates the use of a fleet of trucks to carry cargo directly between the cus-tomer or container freight station and the ship, leading to further delays, greater port congestion (both land- and sea-side) and increased handling costs (Carana Corporation 2004).

The non-working time discussed above is explained in part by the manner in which labour is used at ports, which effectively institution-alizes the underutilization of port facilities and limits the potential for efficiency improvements. In many ports, only one shift of labour is pro-vided and opportunities for overtime are limited. For those ports that are meant to operate on a 24-hour basis, six out of every 24 hours are being lost because of rigid break periods not staggered to ensure con-tinual servicing of vessels (Nathan Associates 2001). Delays attributable to unfairness and corruption in berth assignment are another cause of non-working time. Basri (2005) notes that the use of informal payments to cut queuing time resulting from the lack of key infrastructure facilities such as gantry cranes and storage space is commonplace. Such costs are in addition to a broad range of informal payments required at the port for export and import procedures, as highlighted by media reports.

Costs are also increased by poor port security, with cargo shipments from Indonesia typically attracting an insurance premium that is 30–40 per cent higher than for cargo originating in Singapore. This is explained not only by piracy at sea, but also by the port-based activities of organ-ized crime groups, by more general theft and pilferage and by strikes and work stoppages (Carana Corporation 2004). As noted later, the main ports involved in import–export operations must now upgrade their security to satisfy the new international security requirements of the International Ship and Port Facility Security (ISPS) Code.

## THE 2008 SHIPPING LAW

The new Shipping Law was passed in April 2008 after four years of development. Comprising some 355 articles, the law covers a broad range of maritime-related issues such as shipping, navigation, environ-mental protection, sailor welfare, maritime accidents, human resource development, community involvement and the creation of a coast guard, among many others. The law has received some positive attention in the media, particularly with regard to its provisions on cabotage.[11] The cabo-tage rules, which limit domestic carriage to domestically flagged ves-

---

11  See, for example, 'Bolstering the shipping industry', *Jakarta Post*, 14 April 2008: 6.

sels, represent nothing new for Indonesia and the law essentially restates existing regulations.[12] This focus on cabotage reflects the strength of the local shipping industry lobby, in particular the Indonesian Shipowners Association, which has long advocated the need for a larger domestic cargo fleet to replace foreign-flagged ships on domestic routes.

There is little doubt that Indonesia would benefit from having an upgraded fleet comprising larger and more modern vessels. However, no matter how large and modern the fleet is, the shipping sector will struggle to be profitable if ships must spend a considerable proportion of working time queuing outside, or berthed at, congested ports. In this regard the 2008 Shipping Law is significant in that it provides the foundation for a radical transformation of the national system of port governance that could lead to substantial efficiency improvements in the medium to long term. By removing the IPCs' legislated monopoly on commercial ports, the law opens ports up to participation by other operators, including those from the private sector. It also provides for a clear separation between operator and regulator. Under existing regulations, the IPCs have regulatory authority over other (potentially competing) ports in their respective geographical regions of control. Under the new law, most regulatory authority at the port level will be vested in newly formed port authorities. The role of the IPCs, at least on paper, is subsequently reduced to that of port operator.

This new system of port governance constitutes a common model of port administration known as the 'landlord port'. In simple terms, this model sees the government — as represented by the port authority — own, provide and regulate access to port land and port waters as well as basic port infrastructure such as breakwaters, sea channels and navigational aids. Port operators then lease these facilities and provide port services under a long-term contract or concession (World Bank 2004).

A crucial supporting document is the National Ports Masterplan, which determines both the location and the hierarchy (functions) of current and planned ports.[13] At the port level, port authorities are responsible for individual port masterplans covering such matters as geographic (land and water) working areas, the provision of basic infrastructure and the regulation of port operator access to facilities.[14]

---

12  Such as Government Regulation (PP) No. 17/1988 and Presidential Instruction (Inpres) No. 5/2005.

13  Here 'function' refers to whether the port handles international cargo or is a domestic feeder port.

14  At the time of writing it was still not clear whether there would be a dedicated port authority for each port (comprising multiple terminals) or whether the port authorities would oversee multiple ports.

Like many Indonesian laws, the Shipping Law is very general. Important details will be provided in the implementing regulations which the Ministry of Transport expects to draft by April 2009. Since the passage of the law a number of investors, both local and foreign, have made public their intention to explore new port investment opportunities, most notably the former prime minister of Thailand, Thaksin Shinawatra. However, these investors are unable to pursue their plans until the necessary implementing regulations, planning documents and supporting institutions are developed. Discussed below are five areas where action is required by government such that new investment, and hence competition, can be injected quickly into the ports system.

### Consistency with the negative investment list

The government has long maintained a negative investment list (*daftar negatif investas*, or DNI) with the intention of protecting certain sectors from foreign and/or large investors. In the most recent iteration of the list, issued in December 2007, more industries have been opened up to foreign investment than are closed to it. One exception is the ports sector. According to the latest list, all port activities are now limited to 49 per cent foreign ownership limits. This includes stevedoring, tugboat operation, container terminals, liquid and dry bulk terminals, roll-on/roll-off terminals and investment in wharfs and port superstructure.

The benefits of this restriction, which clearly runs counter to the liberal and pro-competition character of the port section of the Shipping Law, are not immediately evident. Indonesia has an inefficient ports sector suffering from decades of poor governance and underinvestment. Foreign investment would provide crucial upgrades in capacity using best international practices and technology and help inject much needed welfare-enhancing competition into the ports sector. Given that national sovereignty over ports is ensured by the very nature of the landlord port concept, it seems unlikely that the ownership cap is driven by nationalist considerations. Rather, it is more likely that this limitation has been imposed to protect local port operators, both current and future. It is interesting to note that the influential Indonesian Shipowners Association (the largest members of which now have ambitions to become port operators) has publicly urged the government to give priority to local over foreign investors when it implements the port regulations.[15]

---

15 See, for example, the comments by Sungkono Ali, Secretary-General of the Indonesian Shipowners Association, in *Bisnis Indonesia*: 'INSA: utamakan investor local kembangkan pelabuhan' [Indonesian Shipowners Association: prioritize local investors to develop ports], 19 June 2008: R1.

The 49 per cent ownership cap will dissuade some foreign investors from investing in the ports sector. It will make the process more complicated for those who do wish to invest, as they will need to find local partners and explore nominee arrangements. There are considerable economic and political advantages for the major international container port operators in partnering with cashed-up and influential local investors, but the foreign investor would nevertheless insist upon operational control of the port to ensure an adequate return on the investment.

The investment restriction has been criticized openly by the business community and by some within the government.[16] With foreign investors increasingly interested in Indonesian ports, the Ministry of Transport, the key sponsor of the cap, will be under pressure to remove or soften the restriction in subsequent iterations of the DNI.

### The National Ports Masterplan

In mid-2008, the Ministry of Transport began to develop the National Ports Masterplan, a policy document that will determine the location, functions and hierarchy of Indonesia's ports. The plan is expected to be completed by June 2009. The transport minister is responsible for this document, which has a shelf life of 20 years.

Although not articulated in the law, the National Ports Masterplan is expected to give form to the ministry's long-held desire to reduce the number of ports with direct international links. Currently over 100 ports are allowed to have direct international connections. This is expected to be reduced to approximately 25, most probably the 25 strategic ports mentioned earlier.[17]

Port rationalization has its merits. Given the large number of international ports in Indonesia, aggregation could translate into lower per unit cargo-handling and freight costs (Carana Corporation 2004). This of course presupposes competitive feeder services (for transhipped cargo) and the requisite road interfaces (for hinterland cargo). Nevertheless, as discussed earlier, recent international experience suggests that there are considerable efficiency gains to be enjoyed by using larger vessels visiting ports with deeper harbours and more developed cargo-handling infrastructure.

---

16  See, for example, the comments by Mohamad Ikhsan, Expert Staff for the Coordinating Minister for Economic Affairs, in *Media Indonesia*: 'Listrik, pelabuhan sebaiknya tidak masuk DNI' [Electricity, ports should not be on the DNI], 29 July 2008: 17.

17  'Pembatasan pelabuhan terbuka ditetapkan pekan ini' [Restrictions on open ports to be determined this week], *Bisnis Indonesia*, 25 March 2008: R3.

Port rationalization will also make it much easier for Indonesian ports to comply with the ISPS standards developed in the wake of the 11 September 2001 attacks in the United States and the bombing of a French oil tanker in 2002.[18] Until now Indonesia has struggled to meet these standards. In February 2008 the US Coast Guard issued a port security advisory (PSA) for the majority of Indonesia's international ports, whereby ships visiting Indonesian ports within five port calls of the United States must undergo extra security procedures before being allowed to visit US ports. The US Coast Guard has exempted the 16 Indonesian ports that are ISPS-compliant from its PSA requirements. Of those, only eight are commercial public ports (US Embassy Jakarta 2008).

As has been emphasized by the Ministry of Transport, there are two other reasons to pursue port rationalization: to support the implementation of cabotage and to address smuggling.[19] Reducing the number of international gateways will increase the demand for feeder ports, which should benefit domestic shipping companies. In as much as smuggling occurs through international ports, concentrating the customs service and facilities in a smaller number of ports may improve the monitoring of trade flows.[20]

As port rationalization will see some ports and regions lose their direct international links, it is likely to be a contentious issue and to attract considerable debate and scrutiny. On one side of the debate will be producers, exporters and importers from regions that lose their direct international connections, and hence facing higher transport costs. On the other side will be the domestic shipping industry, which will benefit

---

18  The ISPS Code was added to the International Convention for Safety of Life at Sea (SOLAS) in 2002. It represents a comprehensive set of standards designed to improve the security of ships and port facilities. As Indonesia is a signatory to the convention, ISPS standards must be applied to all Indonesian ships of 500 gross tonnes and over on international voyages as well as the ports serving those vessels. It also applies to mobile offshore drilling units.

19  Interviews with various officials. See also the comments by Effendi Batubara, Director-General for Sea Communications, in *Bisnis Indonesia*: 'Cigading diarahkan gantikan peran Pelabuhan Priok [Cigading directed to replace the role of Tanjung Priok], 31 March 2008: R6.

20  Even this reduction in gateways is unlikely to satisfy the Indonesian Textile Association (API), the main association representing the textiles and garments sector. It is a key advocate of stronger measures against smuggled imports (but not against the trade barriers that promote smuggling). API is now recommending only two dedicated ports for textile and garment imports: Tanjung Priok (Jakarta) for western Indonesia and Tanjung Perak (Surabaya) for eastern Indonesia ('Port inspection as culprit in smuggling', *Indonesia Shipping Times*, July 2008: 14).

from the increased demand for domestic feeders. These competing agendas need to be balanced and assessed to ensure the greatest net benefit for the national economy.

Similarly, careful consideration should be given to the possible effects of rationalization on interport competition. Until now, Indonesia has not been able to enjoy the benefits of ports competing in the same hinterland for cargo because of the regulatory and management structures governing the IPCs. Under the new Shipping Law, competition is possible not only within ports (that is, between competing terminals) but also between ports. However, with the development of the National Ports Masterplan, there are concerns that decisions on port location, functions and hierarchy will be made in such a way as to reduce competitive pressures on the incumbent IPCs.

### The role of the port authorities

A key innovation of the new law is the development of port authorities to supervise and manage commercial operations within each port. Their primary responsibility will be to regulate, price and supervise access to basic port infrastructure and services, including port land and waters, navigational tools, pilotage, breakwaters, port basins, sea channels (dredging) and port road networks. In addition, each port authority will be responsible for developing and implementing a port masterplan (including determining land and sea areas of control) as well as ensuring the port's orderliness, security and environmental sustainability. Port operators, meanwhile, will provide cargo-handling and passenger facilities, mooring services, refuelling and water supply, towage, storage and other superstructure.

This is a common division of responsibility across the public and private sectors in a landlord port setting (ADB 2000; World Bank 2001). While there is typically some variation in these arrangements across ports and countries, the general rule is that, where there are public interest or natural monopoly considerations, the functions in question are best provided by government. Indonesia's port authorities will be no exception in this regard and will have roles and functions similar to many port authorities elsewhere. A matter of greater concern is whether Indonesia's port authorities will have the requisite technical and financial capacity to carry out their functions effectively.

Technically, concern will focus on the requirement in the law that only public servants may staff the port authorities (article 86). This is a departure from the recent practice of establishing government regulatory and supervisory bodies (as well as other government agencies that provide key services) and giving them general service agency (*badan layanan umum*, or BLU) status.[21] This type of government legal entity has considerably more flexibility to recruit professional staff. Thus, allowing port authorities to assume BLU status would enable them to recruit higher-paid staff with a more varied skill set, such as retired shippers. Instead, the Ministry of Transport has made it clear that it expects the port authorities to be staffed by a combination of ministry officials from the Sea Communications Directorate and port administration (*administrator pelabuhan*, or Adpel) officials.[22]

The move to a landlord model necessarily means the development of a more complex set of public–private sector interactions at the port level. A crucial task for the port authorities will be to manage those interactions in such a way as to ensure competitive prices and services. However, Indonesia has little experience in managing ports in a competitive context, as the port sector is currently operated by a public sector monopoly characterized by little or no contestability in the provision of services. Where there have been opportunities to introduce competition, these have been poorly managed. Nathan Associates (2001) notes a key example from the late 1990s, where separate concessions for the two main container terminals at the port of Jakarta (JICT and Koja) were sold to the same corporate entity. With the impending deregulation of operator prices (as allowed by the new Shipping Law), the implications of the decision not to sell the concessions to separate and competing entities will become increasingly apparent.

Another issue is how the planned port authorities will interact with the incumbent IPCs. Given the unique historical, institutional and even personal relationships that the IPCs share with the public servants likely to staff the port authorities, there are concerns about possible discriminatory treatment of new investors. This could take many forms, including, for example, unequal access to key facilities and services such as land and basic infrastructure, overly prescriptive and/or restrictive port masterplans that present entry barriers to new investors, discriminatory pricing and so on.

Financially, concern will focus on the ability of the port authorities to fulfil their mandate to provide basic infrastructure. Existing port infrastructure is currently being used by the incumbent IPCs. While some ports may be able to be expanded so that new entrants can use existing

---

21  A recent example of a government agency with BLU status is BP Migas, which was set up to supervise the upstream oil and gas industry.

22  See, for example, the comments by Kholik Kirom, Director of Ports and Dredging, Ministry of Transport, in *Kontan*: 'Undang-undang Pelayaran: otoritas pelabuhan berasal dari pegawai negeri sipil' [Shipping Law: port authorities will be made up of civil servants], 12 May 2008: 13.

facilities such as breakwaters, sea channels and navigational devices, it is also likely that the development of new terminals and operator facilities will require investment in new basic infrastructure. Any delays in such investment will prevent new entry and will obviously advantage the IPCs.

It is therefore critical that the port authorities have the capacity to generate their own sources of funding, and not be entirely dependent on transfers from the central government. Operating expenses, for example, could be financed in part by fees and charges paid by terminal operators, including the fees currently paid by the private (special) ports to the IPCs. New basic infrastructure could be built under concession from the port authorities on some kind of build–operate–transfer basis. However, this would require the development of considerable in-house capacity on the part of the port authorities as well as the necessary approvals from the central government. Moreover, the negative investment list restriction noted above may limit, or at least complicate, foreign investor participation. More ambitiously, port authorities could launch bonds to finance and build basic infrastructure.

### Pricing of port services

According to UNCTAD (1998), the freedom to price according to commercially and financially sound principles is an essential precondition for the successful and sustainable operation of private enterprise within a landlord port context. The same report notes that granting price autonomy to the private operator has four major advantages: it increases the probability that cost-based tariffs will apply (hence improving the chances of a private operator remaining financially viable); it reduces the incentive to pursue cross-subsidization practices (that is, the use of freight rates to cover port costs); it promotes efficiency pricing whereby users who make greater demands pay higher tariffs; and it ensures a stronger link between tariffs imposed and benefits/services provided.

In theory, the new Shipping Law will enable private operators to set their own tariffs. However, the language used in the law raises concern about how much price autonomy operators will actually enjoy. According to article 110(2):

> Port service tariffs will be determined by port operators based on tariff types (*jenis*), structure (*struktur*) and categories (*golongan*) as determined by the government.

This suggests that government will continue to play a role in influencing prices. When interviewed, transport ministry officials insisted that operators would have full price autonomy and that the government would only determine what types of tariffs could be applied and not the allowable tariff levels. It was also apparent, however, that some officials were concerned about the potential under the new model of port governance for 'destructive competition' that might subsequently require government intervention. Further clarification is needed to clearly outline the role (if any) of the government in influencing operator tariffs. Continuing uncertainty in this regard will give rise to concerns that the language of the law will be used to influence prices in such a way as to diminish the competitive advantage of new port operators against the incumbent IPCs.

## The regulation of private (special) terminals

The IPCs have regulatory authority over private ports within the areas they control, and typically use this authority to prevent competition with their own commercial ports. This has been a matter of some policy debate, as many private ports are able to accommodate third-party (general) cargo and have unused capacity. Local governments, empowered by the decentralization process, have been able to challenge the IPCs' authority to regulate private ports to a certain extent (Ray 2003). But for the most part the central government, through the IPCs, has managed to keep a tight rein on private ports, most significantly to prevent competition with the IPCs.

Under the new law, private ports remain tightly regulated and continue to be barred from accommodating third-party cargo. They are no longer referred to as 'ports' but as 'terminals', which are governed by the nearest port authority according to that particular port's masterplan. The law also differentiates between 'special' and 'own-use' terminals. The former are located outside, and the latter within, the port operating area (including both land and waters).

An important change under the new law is that special terminals can apply to become general cargo terminals. This change in status is conditional on approval by the relevant port authority and on being deemed 'consistent' with the local port masterplan, among other requirements. Given the nature of the landlord port model, this change in status will also require all basic port infrastructure (breakwaters, sea channels and so on) to be surrendered to, and then leased back from, the state (as represented by the port authority). This is an important point, as most special terminals by definition are outside the port operating area and have therefore developed their own basic infrastructure.

A study by the Asia Foundation (2008) highlights the critical economic importance of allowing selected private dedicated ports with unused capacity to accommodate third-party cargo. Focusing on Sulawesi, the study notes that most ports on the island are quite small. Very few have a quay length of over 100 metres or an approach draft of over 5 metres, and the draft of most is less than 2.5 metres. This lack of port capacity limits vessel size and hence opportunities for consolidation and bulk handling. Many of the island's main agricultural commodities, such as cocoa, are shipped in bags, at a relatively high cost of around US$165 per tonne to Europe. If these same commodities were loaded and shipped in bulk, the rate to Europe would fall to around US$80 per tonne and the required vessel time in port would fall from six to two days.

The study identifies a private (foreign-owned) dry bulk terminal in the port of Makassar handling mainly wheat that could accommodate large dry bulk vessels. It is a well-equipped terminal with over 250 metres of wharf length capable of handling vessels of 60,000–80,000 dead-weight tonnes (DWT). In addition, it has existing (or soon to be installed) bulk loading/unloading equipment that could easily be converted or upgraded to handle other important dry bulk commodities from the hinterland, such as cocoa. Also, because it is located inside the port of Makassar, it already has access to basic infrastructure. Most importantly, it is only operating at around 20 per cent capacity and the owners are interested in selling this unused capacity to other users.

This terminal represents just one of a number of opportunities for port status conversion in Sulawesi identified by the Asia Foundation; many others can be found in other parts of the country. The best known example is the port of Cigading in Cilegon, Banten, which is owned and operated by PT Krakatau Steel. It is the largest and deepest dry bulk terminal in Indonesia, capable of handling large vessels of up to 150,000 DWT. Only 30 per cent of this capacity is required for Krakatau Steel's own purposes. The port is currently applying to change its status from a special to a commercial port, but it has openly serviced third-party cargo for years.[23]

## OPTIONS FOR PROMOTING COMPETITION AND PRIVATE SECTOR PARTICIPATION

The new governance structure to be established by the 2008 Shipping Law provides three avenues to promote competition and private sector participation in Indonesian ports. The first involves the unbundling of existing port assets such that they can be broken into separate, and preferably privately held, competing entities. This is a favoured option in the

---

23 'Cigading diarahkan gantikan peran Pelabuhan Priok [Cigading directed to replace the role of Tanjung Priok], *Bisnis Indonesia*, 31 March 2008: R6.

privatization literature to achieve the immediate injection of competition into infrastructure sectors hitherto dominated by a state monopoly. However, in this case it is probably the most politically difficult option to pursue. As noted in a variety of media reports in the weeks leading up to the enactment of the Shipping Law in April 2008, there was considerable opposition to the law from port unions, which were threatening strike action. More subtle opposition came from the IPCs and the Ministry of State-owned Enterprises. In response, the government has made a clear commitment that no major IPC assets will be sold to the private sector.

The second option is greenfield investment in new terminals. This provides an important mechanism to upgrade capacity and enhance competition in the medium to long term. However, it would require lifting (or at least softening) the foreign investment cap on port operations and the development of new basic infrastructure by the government, as well as a host of regulatory approvals, all of which is likely to take time. Most importantly, it would require the establishment and subsequent capacity development of a number of civil servant-manned port authorities to oversee planning and port operations and regulate access to key port services and facilities. This too would take time, and new investors would be cautious as to how they would be treated by the new authorities compared with their incumbent IPC competitors.

The third, and perhaps most feasible, option to promote immediate competition and private sector participation in Indonesian ports is to allow the conversion of special and own-purpose private terminals such that they can accommodate general cargo. Indonesia currently has considerable unused container and bulk-handling capacity in these private ports that could be used to provide competition for the IPCs. Allowing at least some of them to accommodate third-party cargo would provide some short to medium-term solutions for Indonesia's current port logistics problems. Longer-term solutions, meanwhile, must wait for the investment in new capacity made possible by the 2008 Shipping Law.

## REFERENCES

ADB (Asian Development Bank) (2000), *Developing Best Practices for Promoting Private Sector Investment in Infrastructure: Ports*, Manila.

Asia Foundation (2008), *The Opportunities for Improving Port Efficiency in Sulawesi Created by the New Port & Shipping Law*, Jakarta, August.

Basri, M. Chatib (2005), 'Competitiveness of Indonesian industries from logistics perspective: inefficiency in the logistics of export industries', Institute for Economic and Social Research, Faculty of Economics, University of Indonesia (LPEM-FEUI), Jakarta.

Carana Corporation (2004), 'Impact of transport and logistics on Indonesia's trade competitiveness', report prepared for the USAID-funded Trade Enhancement for the Services Sector (TESS) Project, Washington DC, June.

ESCAP (Economic and Social Commission for Asia and the Pacific) (2007), *Regional Shipping and Port Development: Container Traffic Forecast, 2007 Update*, ESCAP and Korean Maritime Institute, United Nations, New York.

Kruk, C. (2008), 'State of the port sector', report presented to the World Bank Roundtable on Logistics, Jakarta, 19 June.

Ministry of Transport (2006), *Buku Informasi 25 Pelabuhan Strategis Indonesia* [Information Book on the 25 Strategic Ports of Indonesia], Direktorat Pelabuhan dan Pengerukan, Jakarta.

Nathan Associates (2001), 'Indonesia shipping and port sector review', technical report prepared for the USAID-funded Partnership for Economic Growth Project, Jakarta.

Patunru, A.A., N. Nurridzki and Rivayani (2007), 'Port competitiveness: a case study of Indonesia', LPEM–FEUI and Asian Development Bank Institute, available at http://www.adbi.org/files/2007.06.26.indonesia.sea.ports.competitiveness.pdf.

PDP Australia (2005), 'Promoting efficient and competitive intra-ASEAN shipping services: Indonesia country report', REPSF Project No. 04/001, report produced by PDP Australia Pty Ltd and Meyrick and Associates for the ASEAN Secretariat and AusAID, March.

Penfold, A. (2007), 'Trade concentration and the use of large vessels in the container trades', paper presented to the twelfth Congreso de Trafico Maritimo, La Caruna, April.

Ray, D. (2003), 'Survey of recent developments', *Bulletin of Indonesian Economic Studies*, 39(3): 245–70.

UNCTAD (United Nations Conference on Trade and Development) (1998), *Guidelines for Port Authorities and Governments on the Privatization of Port Facilities*, Geneva.

US Embassy Jakarta (2008), 'US Coast Guard issues advisory to Indonesia on port security', press release, Public Affairs Section, 26 February.

World Bank (2001), *Port Reform Toolkit*, Public–Private Infrastructure Advisory Facility, Washington DC.

World Bank (2004), *Reforming Infrastructure: Privatization, Regulation and Competition*, Washington DC.

# 7 PIRACY AND ARMED ROBBERY AGAINST SHIPS IN INDONESIAN WATERS

*Sam Bateman*

---

The Indonesian archipelago sits astride key shipping routes that carry a large share of the world's seaborne trade between the Indian and Pacific oceans. Most international interest in maritime security in recent years has focused on the Malacca and Singapore straits, but shipping elsewhere in the Indonesian archipelago, particularly that passing through the Lombok and Makassar straits, is potentially vulnerable to attack. It is ironic that while most interest has focused on the vessels passing through or near the Indonesian archipelago, the real problem of security is the more grassroots one of protecting vessels trading locally, or those in port or at anchor (Bateman, Ho and Mathai 2007: 311).

As international attention on piracy has shifted to the waters off Somalia, Indonesia no longer has the 'most pirate-infested waters in the world' (Storey 2008: 106). The threat of piracy and armed robbery against ships in Indonesian waters has declined considerably in recent years.[1] From 121 incidents reported by the International Maritime Bureau (IMB) in 2003, accounting for about 27 per cent of the world total, the annual number of incidents in Indonesian waters fell steadily to 28 incidents, or less than 9 per cent of the world total, in 2008 (IMB 2009: 5, Table 1). In

---

1 This chapter follows the International Maritime Organization in using the expression 'piracy and armed robbery against ships' to describe incidents, in order to overcome the limitation under international law that an act of piracy can occur only in international waters. This expression covers not only acts against vessels at sea, but also acts against vessels in port or at anchor, regardless of whether they were inside or outside the territorial sea or archipelagic waters when attacked.

2003, Southeast Asia was the most piracy-prone area in the world, with 189 incidents out of a total of 445 actual and attempted attacks worldwide. The IMB annual report for 2003 noted that the north Sumatra/Aceh coast was 'particularly risky', with numerous violent attacks reported during the year (IMB 2004: 15).

There are several reasons for the improvement. First, the decline in the number of attacks around Sumatra and adjacent areas of the Malacca Strait appears to be largely due to the improved political situation onshore following the peace agreement in 2005 between the Indonesian government and the Free Aceh Movement (Gerakan Aceh Merdeka, or GAM). The fall in the number of attacks off northern Sumatra, particularly off the port of Belawan near Medan, has confirmed suspicions that elements of GAM were involved in piracy and sea robbery. On the basis of extensive research on attacks on fishing vessels attributed to the key fishing community of Hutan Melintang on the west coast of peninsular Malaysia, Malaysian maritime security analyst Mak Joon Num asserts that until 2005 members of GAM were heavily involved in organized piracy, including a maritime protection syndicate. Since then, piracy among the Hutan Melintang fishers has ceased completely (Mak 2007, 2008: 4). The December 2004 tsunami also contributed to the fall in the number of attacks off Sumatra, as many small craft were destroyed by the tsunami and the 'pirates' lost their means of attacking ships. Some analysts claim that the presence of foreign naval vessels on disaster relief operations in the area was another factor in reducing the number of attacks (Raymond 2007: 65), but the accuracy of these claims is difficult to assess. Most of these naval operations were focused on the northwestern side of Sumatra, whereas most incidents of piracy and sea robbery occurred off the east coast and in the Malacca Strait.

More extensive national and regional responses to the problems of piracy and armed robbery at sea are the second major factor leading to a reduction in attacks. The broadest regional mechanism is the Regional Cooperation Agreement on Combating Piracy and Armed Robbery against Ships in Asia (ReCAAP), which became operational in September 2006 with the opening of its Information Sharing Centre in Singapore. This Japanese-inspired initiative was a very significant achievement in that it provided the basis for regional cooperation to counter piracy and sea robbery. ReCAAP's information network and cooperation regime, with responsible agencies as contact points in each participating country, comprises Japan, China, Korea, India, Bangladesh, Sri Lanka and all ASEAN nations except Indonesia and Malaysia.

The reluctance of Indonesia to join ReCAAP stems from its concern that the organization undermines the concept of state sovereignty in archipelagic waters and territorial seas (Bradford 2008: 489). Similar concerns underpin Indonesia's reluctance to ratify the 1988 Convention for the Suppression of Unlawful Acts against the Safety of Maritime Navigation (the SUA Convention), even though it expands the scope for an international regime to act against piracy and sea robbery. As amended by its 2005 Protocol, the SUA Convention overcomes the limitation of the anti-piracy regime in the 1982 United Nations Convention on the Law of the Sea (UNCLOS), under which only acts taking place in international waters can be treated as piracy (Beckman 2008: 188–92). Malaysia's reluctance to support ReCAAP lies in its objections to the location of the Information Sharing Centre in Singapore. Malaysia regards ReCAAP as an unnecessary competitor to the IMB's Piracy Reporting Centre, which is located in Kuala Lumpur. Both Indonesia and Malaysia, however, have signalled a willingness to cooperate with ReCAAP (Raymond and Morrien 2008: 8–9).

Increased patrolling and surveillance in the Malacca and Singapore straits has also contributed significantly to the reduction in piracy and armed attacks at sea. Indonesia, Malaysia and Singapore conduct trilateral, coordinated maritime surface patrols, known as the Malacca Strait Sea Patrols, and coordinated airborne surveillance under the 'Eyes in the Sky' arrangement (Bradford 2008: 482). Other forms of cooperation between the littoral states include an agreement between Malaysia and Indonesia in 2007 to increase joint anti-piracy training in the Malacca Strait,[2] the Surface Picture Surveillance System (SURPIC) launched by Singapore and Indonesia in May 2005 to provide real-time surveillance of the Singapore Strait and its approaches, and the Malacca Strait Patrol Information System (MSP-IS) to share information about shipping in the Malacca Strait (Ho 2009: 3). These operations mainly have a deterrent effect, as few pirates are actually caught at sea.

The Malacca and Singapore straits are now covered by the Cooperative Mechanism for the Straits of Malacca and Singapore, established by the International Maritime Organization (IMO) in September 2007 to encourage user states and other stakeholders to cooperate voluntarily with Indonesia, Malaysia and Singapore in enhancing safety, security and environmental protection in the straits (Sasakawa 2007). It comprises three elements: a Cooperation Forum, an Aids to Navigation Fund and specific projects. The Cooperation Forum brings together the littoral states, user states and other interested stakeholders to discuss matters of mutual interest. The Aids to Navigation Fund is funded by the voluntary contributions of user states and other stakeholders. Its purpose is to

---

2 'Malaysia, Indonesia to increase joint anti-piracy training in the Malacca Strait', *Bernama*, 13 June 2007, reported by BBC Monitoring Asia Pacific, 14 June 2007.

enhance navigational safety and environmental protection in the straits by funding the purchase and maintainance of such navigational aids as lighthouses and buoys. So far, however, security issues have not been a direct concern of the Cooperative Mechanism (Ho 2006).

At a national level, tighter government control and better local policing onshore have helped to reduce attacks. As the rise of piracy and sea robbery off the coast of Somalia demonstrates, corrupt governance or a lack of good order onshore facilitates disorder at sea (Murphy 2007: 15; Storey 2008: 106). Pirates operate from bases onshore, usually in small fishing communities where it is not unreasonable to assume that most of the community knows what is going on. This includes local police and naval personnel, who may even be complicit in the illegal activity. This appears to have been the case in Indonesia in the past. Eric Frécon, who has conducted extensive field research on piracy in the Riau Archipelago, says that there is little doubt 'that local policemen are fully aware of the criminal activities that are carried on by some of their neighbours', and that 'Sometimes policemen are not only tolerant but also accomplices' (Frécon 2008: 10). Low salaries for law enforcement personnel encourage such complicity.

Current developments within Indonesia include the establishment of a Sea and Coast Guard — Kesatuan Penjaga Laut dan Pantai, or KPLP — under the Ministry of Transport. The Maritime Security Coordinating Board (Badan Koordinasi Keamanan Laut, or Bakorkamla), meanwhile, has broad oversight of sea security and relevant policy. Indonesia's national efforts have been hampered both by a lack of capacity to conduct security operations and by a lack of coordination between the 12 government agencies that have responsibility for some aspect of maritime law enforcement (Dirhamsyah 2005: 3, Table 1). The situation is further complicated by the country's regional autonomy laws, which devolve some responsibility for law enforcement at sea to provincial and local governments. Hitherto the responsibility for all law enforcement at sea rested largely with the Indonesian navy, which had the ships and aircraft but not the legal powers to deal with the range of illegal activities at sea. The navy jealously guards its offshore role; it does not want to surrender its facilities or other interests to the coast guard. In spite of the fact that the coast guard has good, well-maintained, high-tech patrol boats acquired from Germany, it cannot obtain the funding it needs. As a result, its main role appears to be limited to investigating issues such as marine pollution, the theft of oil, drug smuggling and theft (especially of navigational aids) from vessels.

Several countries have provided assistance to improve security in Indonesian waters. Japan has donated new patrol vessels and the United States has provided funding and technical assistance to establish 12 surveillance radars along the Malacca Strait and a further eight along the Makassar Strait.[3] Scheduled to become operational in 2009, the Malacca Strait radars are to be manned by the navy and operated centrally from Batam[4] as part of a planned Integrated Maritime Surveillance System (Hammick 2007: 3). China has also announced that it will give Indonesia a remote sensing satellite to monitor activities at sea.[5] While this assistance with surveillance systems may be advantageous politically, particularly for the donor countries, great challenges are involved in linking the systems and making them operationally effective. The problems are more serious because the systems are arriving from disparate sources with no coordination between the donor countries.

Official and community attitudes towards piracy in Indonesia have nevertheless hardened significantly in recent years (Bradford 2008: 480). As Frécon (2008: 13) observes, 'piracy is no longer in fashion in the South of the Malacca Straits'. Word has gone out in Indonesia that acts of piracy and sea robbery are unacceptable because they have the potential to harm the image of the country. In addition, in most of the piracy-prone areas, police are more active and better equipped than previously. This environment has forced pirates to retire, pursue other illegal activities, notably smuggling, or shift to areas where there is less of a security presence, such as the area off Mangkai in the Anambas Islands in the South China Sea (Frécon 2009: 2).

The final factor in the decline in the number of incidents of piracy and armed robbery against ships in Indonesian waters is the change that has occurred since 11 September 2001 in the general attitude of the international shipping industry towards maritime security. Following the introduction of the International Ship and Port Facility Security (ISPS) Code by the IMO in 2002, all ships above a certain size engaged in international voyages are required to have a range of additional security measures in place, including a nominated ship security officer and a ship security plan.[6] They must also have a security system fitted to transmit an alert in the event of an emergency onboard, although doubts have arisen as

3 'Indonesia: surveillance radar provided by US to be erected in Tarakan, East Kalimantan', *Media Indonesia*, 16 October 2008, reported by BBC Monitoring Asia Pacific, 17 October 2008.

4 'Indonesia to make 12 Malacca Strait surveillance radars operational soon', 6 January 2009, reported by BBC Monitoring Asia Pacific, accessed 12 January 2009.

5 'China to grant remote sensing satellite to Indonesia', *Antara News*, 22 January 2009, available at http://www.anyatara.co.id/en/print/?i=1232581524, accessed 23 January 2009.

6 For a comprehensive discussion of these new maritime security requirements, see Jones (2008).

to the effectiveness of the system.[7] These developments have led to a higher level of vigilance and concern for security onboard ships than was previously the case. New seafarer identification documentation has also made it more difficult for pirate gangs to infiltrate the crews of ships targeted for attack (Tsamenyi, Palma and Schofield 2008). Other important new measures include a requirement for ships to have a specific ship identification number and continuous synoptic record, which are routinely inspected and verified by port authorities.[8] These measures make it extremely difficult to hijack a ship and give it a false identity either to offload its cargo or to use it for further trading.[9] No such incidents have occurred in Southeast Asia since the ISPS Code and associated measures were introduced. The small tugs that are still occasionally hijacked are not required to conform to these measures.

## ONGOING CHALLENGES

Although the overall situation with piracy and armed robbery against ships in Indonesian waters has improved in recent years, incidents still occur regularly. A review of the actual and attempted attacks in 2008 provides the basis for identifying patterns in these attacks, current 'hot spots' and recommendations on how the current situation might be improved and piracy in Indonesian waters eradicated.

Table 7.1 summarizes the 26 actual and two attempted attacks in Indonesian waters during 2008. It is fair to assume that these figures under-

---

7    A study conducted by the S. Rajaratnam School of International Studies in Singapore found that long delays were experienced in an alert reaching a responsible authority, and that the system was plagued by a high number of false alerts (Timlen 2008).

8    A continuous synoptic record provides a verifiable account of a ship's history, including any changes of name, ownership or flag. It would be extremely difficult to create a 'phantom ship' with a continuous synoptic record that would survive even the most rudimentary check of its authenticity.

9    The hijacking of the Singapore-flagged product tanker *Petro Ranger* in April 1998 while on passage from Singapore to Ho Chi Minh City is an often quoted example of a 'phantom ship'. After leaving Singapore, the ship was boarded by Indonesian pirates in the southern part of the South China Sea. The *Petro Ranger* was then given a false identity under a flag of convenience as the M.T. *Wilby*, with fraudulent registration papers apparently prepared in advance. Some of the ship's cargo of automotive diesel distillate and aviation fuel was discharged into smaller tankers before the *Wilby* was taken into the Chinese port of Haikou, where the remaining cargo was discharged. After prolonged investigation by the Chinese authorities, the crew of the *Petro Ranger* was released and the ship returned to its rightful owners. The full story of this incident is available in Blyth and Corris (2000).

*Table 7.1    Attacks on vessels in Indonesian waters, January–September 2008*

|  | Actual | Attempted |
|---|---|---|
| Type of attack | 25 boarded<br>1 hijacked | 1 fired upon<br>1 boat approach |
| Situation of vessel | 2 berthed<br>13 anchored<br>11 steaming | 2 steaming |
| Type of vessel | 6 bulk carriers<br>6 chemical tankers<br>3 product tankers<br>2 general cargo ship<br>2 container ships<br>3 LPG tankers<br>1 vehicle carrier<br>1 tug<br>2 barges | 1 bulk carrier<br>1 livestock carrier |
| Level of attack | 2 medium level<br>24 low level | 2 low level |
| Type of weapons | 12 knives<br>3 guns<br>11 not stated | 1 guns<br>1 not stated |
| Type of violence | 16 taken hostage (*Blue Ocean 7*)<br>6 missing (*Blue Ocean 7*)<br>2 injured | |

*Source:* IMB (2009).

estimate the actual number of attacks by Indonesian pirates, because some attacks would not have been reported to the IMB (Chalk 2008: 7). More importantly, however, some or all of the attacks reported in the Malacca and Singapore straits and near Pulau Tioman off the east coast of Malaysia most probably were carried out by Indonesian pirates, who are known to move outside Indonesian waters to find their targets. For example, between September 2008 and November 2008, four incidents of armed robbery took place in the waters off Tanjung Ayam in Johor, all involving vessels at anchor (ReCAAP 2008b: 15). There have also been attacks on vessels passing through or near the east-bound separation lane of the Philip Channel, which lies partly within Indonesian waters,[10]

---

10  There were three attacks in this area in October 2008 with a further two in November 2008 (ReCAAP 2008b: 13–15).

as well as in the western approaches to the Singapore Strait. 'Hit and run' opportunity attacks were fairly common there until recent years.

Bulk carriers form the largest category of ships in Table 7.1, with six actual attacks and one attempted attack. Three of these were on vessels under way and four on ships at anchor. There are three factors that help explain why bulk carriers, regardless of size, are the class of vessel most frequently attacked while under way. First, bulk carriers are relatively slow and have a low freeboard when laden. Second, they are relatively unsophisticated in build and technology and generally have lower standards of ship maintenance and crew proficiency than ships with higher-value cargoes, such as tankers and container ships. This suggests they may be less vigilant and security conscious even when in piracy-prone areas. The third factor concerns the way in which bulk carriers are employed. Analysis of ship records shows that, at the time they came under attack, many of these ships were travelling very slowly, waiting for a new spot charter[11] or preparing to enter port for maintenance or bunkering. This made them easier to approach by pirates.

The distinction between ships attacked while in port or at anchor and those attacked while steaming is important. The majority of attacks in Southeast Asia continue to be on vessels at anchor, in port, or entering or leaving a harbour. The vulnerability of vessels in port or at anchor is not dependent on their size or type. Any vessel in port or at anchor may be attacked, particularly in and around ports with a high incidence of attacks, if full precautions are not taken. Of the 65 actual and attempted attacks in Southeast Asia in 2008, 35 were on vessels that were not at sea (IMB 2009: Tables 4 and 5). Such attacks are usually of a minor nature; they are best countered by more effective policing by port authorities, including active patrolling of ports and anchorages. The successful attacks that do occur at sea on vessels under way are mostly on small vessels and vessels that have slowed down for some reason. Most large, modern merchant ships engaged in international trade travel at speeds in excess of 14 knots, making it both difficult and dangerous for small craft to attempt to approach them. However, attacks on vessels under way tend to be more significant in terms of the level of violence used, the number of attackers and the economic loss suffered (ReCAAP 2008d: 10).

The largest vessel attacked in Indonesian waters in 2008 was the Panamanian-flagged bulk carrier *Princess Nadia*, weighing 85,000 gross

---

11    A spot charter is a charter for a particular vessel to move a single cargo between specified loading and discharge ports in the immediate future. Bulk carriers are the modern-day tramps of the sea; they are often chartered for a single voyage rather than several voyages, and sometimes loiter at sea waiting for a new spot charter.

register tonnes (GRT).[12] It was boarded on 29 January 2008 by four robbers armed with knives while the ship was at anchor off the Pulau Laut coal terminal in the southeastern corner of Kalimantan (IMB 2009: 39). The robbers were detected but managed to steal some ship stores before escaping in a speedboat.

The most serious incident in Indonesian waters in 2008 was the hijacking of the *Blue Ocean 7* while on passage from Sulawesi to Surabaya with a cargo of palm oil (IMB 2009: 42-3). It was attacked on 21 May 2008 by about 10 pirates armed with guns and knives. The *Blue Ocean 7* was a small, Indonesian-flagged product tanker on a domestic voyage, and thus not required to be compliant with the ISPS Code. The crew were taken hostage and the cargo discharged at an unknown destination. The vessel was recovered by the Royal Malaysia Marine Police on 19 June 2008 in Sandakan. While six of the crew are still listed as 'missing', this more likely means that they have not been reported to the authorities rather than that they are dead. They may even have been complicit in the attack.

Of the 11 vessels attacked while steaming (IMB 2009: Table 4), most were relatively small vessels. The exceptions were the 25,892 GRT, Norwegian-flagged bulk carrier *Spar Cetus*, attacked off Mangkai Island on 13 April 2008, and the 15,099 GRT, Panamanian-flagged bulk carrier *J.K.M. Muhieddine*, attacked near the Anambas Islands on 30 September 2008. The exact circumstances of these attacks are not known, but it is possible that the vessels were loitering in the area rather than steaming through it. The other nine vessels attacked while steaming comprised one small tug and barge, one barge, one tug, two small product tankers, one LPG carrier, one container ship and one small chemical tanker. The number of small tugs attacked in the waters around Indonesia, including the South China Sea and the Malacca and Singapore straits, was higher in 2008 than in previous years (ReCAAP 2008d: 18).

This analysis relies mainly on data from the IMB, which need some interpretation. On the one hand, there may be underreporting of attacks (Bradford 2008: 475). Both the IMB and the IMO have noted the reluctance by some shipmasters and shipowners to report minor incidents due to concerns that any investigation might disrupt the ship's schedule and that insurance premiums might increase. Underreporting may also occur because attacks on local craft, such as fishing boats, barges and small

---

12 Gross register tonnes (GRT) is a measure of the total internal (enclosed) volume of the vessel in units of 100 cubic feet (equal to one tonne) less the volume of certain exempted spaces catering for the crew's comfort and the ship's safety. It is to be distinguished from deadweight tonnes (DWT), which is a measure of the total weight of the cargo, stores, fuel and water needed to submerge a ship from its light draught to its maximum permitted draught.

barter vessels, may not be reported to the IMB (Mak 2008: 4). However, there is also the risk of overreporting. The widespread use of the internet makes it very easy to report an attack by email, and many attacks, particularly ones of a minor nature, that may not have been reported in earlier years are now being reported. This is borne out to some extent by the relatively greater number of minor and attempted attacks appearing in the IMB statistics over the last 10 years (Bateman, Raymond and Ho 2006: 20).

It should be noted also that there is some inconsistency between the IMB data and the data collected by ReCAAP's Information Sharing Centre. The latter reported 25 attacks (24 actual and one attempted) in Indonesian waters in 2008 (ReCAAP 2008d: 8), whereas the IMB reported 28 attacks (26 actual and two attempted). The discrepancy may arise from the different approaches the two organizations take to categorizing attacks. Indonesian authorities have argued over the years that the IMB statistics overstate the piracy problem in Indonesian waters. Although this claim is hard to test, there does appear to be a tendency for the IMB to overdramatize certain incidents. According to the IMB, for instance, the feeder container ship *Sinar Merak* was attacked on 22 January 2007 in the Malacca Strait after leaving Belawan for Singapore. However, subsequent investigation by the Singapore security agencies revealed that two suspected pirates found onboard the *Sinar Merak* were actually innocent Indonesian fishermen whose small craft had been run down by the container ship as it manoeuvred aggressively to avoid an attack (ReCAAP 2007: 11). Nevertheless, the IMB continues to show this incident as an actual attack in the Malacca Strait.

The IMB's categorization of attacks by vessel type is also unsatisfactory for making detailed assessments of the risks faced by different types of ships (Bateman, Ho and Mathai 2007: 314). While 37 different types of ship are listed in the IMB database, many are specialist vessel types (such as cable layers or dredgers) that record very few attacks. On the other hand, some major categories (especially the container ship, bulk carrier and product tanker categories) include many ships of vastly different sizes and purposes. For example, a 'product tanker' may be a large vessel of 50,000 GRT carrying refined petroleum products or a very small vessel of less than 1,000 GRT carrying palm oil. Because this category is so broad, it is difficult to assess which kinds of vessels are most at risk.

## HOT SPOTS

There are several regional hot spots in and around Indonesia where incidents of piracy and armed robbery against ships still occur. Coincidentally, these are also the areas where maritime boundaries have not been completed and maritime enforcement is problematic. Parts of the Malacca Strait, the Riau Archipelago, the southern part of the South China Sea and the triborder area in the Sulu and Sulawesi seas are of particular concern. The IMB currently identifies three piracy-prone areas in Indonesian waters: around the Anambas and Natuna islands; near the port of Belawan in Sumatra; and near Jakarta (IMB 2009: 23). In December 2008, the IMB also warned ships to watch for 'suspected Indonesian pirates' off Pulau Tioman following four attacks in this area.[13] Lloyd's Joint War Committee (2009) lists the ports of Balikpapan, Jakarta and Poso (Sulawesi), the northeast coast of Sumatra, the northeast coast of Borneo between Kendat and Tarakan, and the island of Ambon as at-risk areas.

The Riau Archipelago to the south of Singapore, including the major islands of Bintan and Batam, continues to rate as a regional hot spot, with some upsurge in activity during 2008. Historically the area is one of the most piracy-prone in Southeast Asia (Eklöf 2006: 44). Attacks tend to be hit and run, mainly under the cover of darkness, with parangs and pistols as the chief weapons. The pirates are usually satisfied if they can gain access to the ship's safe, seize any valuables and rob the crew. A high level of patrolling and enforcement by security forces, particularly Singapore, has reduced the number of attacks in the area. However, this is not to say that the area is now law abiding. Other forms of maritime crime are prevalent in the region, including extortion and smuggling, particularly of drugs and people. The robbery attacks that still occur at sea largely target small vessels and ships at anchor.

Maritime enforcement in the northern Malacca Strait, directed particularly against illegal fishing, is complicated by the lack of an agreed exclusive economic zone (EEZ) boundary between Indonesia and Malaysia. The agreed continental shelf boundary in this area is to the west of the median line, that is, closer to Sumatra. Indonesia believes that the EEZ boundary should be the median line, whereas Malaysia believes that it should coincide with the continental shelf boundary. When Malaysian fishermen, who are better equipped with larger boats than their Indonesian counterparts, attempt to fish in the disputed area, they may be exposed to enforcement action by Indonesian marine police or naval personnel. This is a possible explanation for reports that 'rogue Indonesian officials' continue to seize fishing boats in the Malacca Strait and demand ransoms for their release (Mak 2008: 6).

There were reports in December 2008 of 'pirates in military-style uniforms' firing on Malaysian fishing boats in the southern Malacca Strait

---

13 'Piracy alert off Pulau Tioman', *Star*, 8 December 2008, available at http://thestar.com.my/news/story.asp?file=/2008/12/8/nation/2747892&iec=nation, accessed 12 January 2009.

off Johor.[14] They were said to be robbing fishermen of their equipment and demanding protection money. As in the northern part of the strait, this new development has some of the indications of rogue enforcement against alleged illegal fishing. It may also be occurring in an area in the western approaches to the Singapore Strait where territorial sea boundaries have not been agreed.

A significant number of attacks also occurs in the southernmost parts of the South China Sea near Pulau Tioman off the east coast of Malaysia, and near the Anambas and Natuna islands in Indonesia. The former are listed by the IMB as having occurred in Malaysia and the latter as having occurred in Indonesia. Table 7.2 lists the attacks in these areas during 2008. As the two areas are not much more than 100 nautical miles apart, it is not inconceivable that the attacks were carried out by the same group of pirates, probably from Indonesia. Most of the attacks involved robbers armed with knives and machetes boarding the vessels and stealing personal valuables, cash and ship's property. The tug *Whale 7* and its barge, however, were hijacked and their crews put ashore. The vessels were subsequently recovered (ReCAAP 2008c).

Frécon (2009: 3) speculates that the reasons for the increase in the number of attacks off Mangkai and other islands in the Anambas group lie in their remoteness, with a limited police and naval presence. He also canvasses the possibility that the perpetrators come from Thailand, in view of the fact that some of the targeted vessels were en route to Thailand.

Finally, the Sulu and Sulawesi seas are a regional maritime-security hot spot because of the presence of terrorists and separatist movements in the area; because of the lack of agreed maritime boundaries and the apparent difficulties in establishing an effective cooperative regime for maritime surveillance and enforcement; and because of the relatively high level of maritime crime in the area, including maritime extortion and the smuggling of goods and people. Security cooperation in this area is said to be 'seriously lacking' (Storey 2008: 120). Extremist groups, including the Abu Sayyaf Group, are active in the area. Piracy, extortion and other forms of maritime banditry have a long history in the Sulu archipelago (Eklöf 2006: 9–13). Many minor attacks in this area on fishermen, small passenger craft and barter traders probably go unreported.

A noteworthy attack in this area was that attempted on the livestock carrier *Hereford Express* on 7 June 2008 in the vicinity of the Sangir Islands in the eastern Sulawesi Sea (IMB 2009: 67). Pirates in a speedboat inter-

---

14 'Malaysia seeks Indonesian help against military-style pirates', *Agent France Presse*, 30 December 2008, available at http://ca.news.yahoo.com/s/afp/081230/world/malaysia_indonesia_piracy, accessed 1 January 2009.

Table 7.2    Actual attacks on vessels under way in the southern area of the
South China Sea, 2008

| Date | Ship's name | Vessel type | Location |
| --- | --- | --- | --- |
| 24 March | *Ocean Seal* | Lift barge | Off Anambas Island |
| 13 April | *Monalisa* | Product tanker | Off Mangkai Island |
| 13 April | *Spar Cetus* | Bulk carrier | Off Mangkai Island |
| 25 April | *Pataravin 2* | Product tanker | Northeast of Singapore Strait |
| 30 April | *PU2008* *PU3306* | Tug Barge | Off Tioman Island |
| 3 June | *Medbothian* | Container ship | Off Anambas Island |
| 4 June | *Red Wing* | Chemical tanker | Off Anambas Island |
| 29 June | *Wecoy 6* | Tug | Off Tioman Island |
| 7 September | *Whale 7* *Sinobest 2503* | Tug Barge | Off Tioman Island |
| 30 September | *J.K.M. Muhieddine* | Bulk carrier | Off Anambas Island |
| 2 October | *Sun Geranium* | Chemical tanker | Northeast of Tioman Island |
| 3 October | *Diamond Coral* | LPG tanker | Off Mangkai Island |
| 1 December | *Entebe Star 21* | Tug (with barge) | Off Tioman Island |

*Source:* IMB (2009); ReCAAP (2008a).

cepted the ship, which was en route to Australia, and opened fire on it. After two hours of continual firing, evasive manoeuvres by the ship were successful and the pirates abandoned their efforts to board the ship. The *Hereford Express* then returned to General Santos in the Philippines with serious damage to its bridge, communications equipment and hull. This attack happened in the same area where an earlier attack, also unsuccessful, had been made on the Indian bulk carrier *Murshidad* on 4 March 2008.

## CONCLUSIONS

Although vessels travelling in Indonesian waters are vulnerable to attack, the risks can be overstated, particularly in terms of the types of ships that are threatened. An accurate risk assessment requires closer analysis of the types of ships that are attacked, their location and their employment.

High-value traffic proceeding through the Indonesian archipelago is normally not attacked. The vessels at greatest risk are ships at anchor or at berth in particular ports; small vessels that are passing through piracy-prone areas; and larger vessels that slow down or loiter in such areas. A situation like that in Somalia, where vessels need to be escorted by foreign warships, is inconceivable in Southeast Asian waters, because pirates lack the strong land base that is necessary for operations against large vessels under way at sea (Bateman and Ho 2008). Under these circumstances, Southeast Asian countries generally, and Indonesia in particular, remain strongly opposed to outside assistance in providing maritime security in the region (Huang 2008).

The main challenge that Indonesia faces if it is to further reduce the risks is to enhance security in particular ports and anchorages and provide greater security in the areas where attacks are still occurring on vessels under way. Improving the security of ports is the responsibility of port authorities, although a greater level of international assistance with building the capacity of these authorities would be helpful. Better coordination between the various agencies involved in providing security in and around ports is required.

Indonesia has a sovereign responsibility to provide security in its archipelagic waters and territorial seas. However, as all its waters abut the waters of a neighbouring country, enhanced cooperation between adjacent countries is essential. Agreement on maritime boundaries in the region is a key requirement, as the lack of boundaries inhibits cooperation. The multilateral cooperation that is occurring in the Malacca and Singapore straits provides a model for areas where cooperation is currently lacking, particularly the southern part of the South China Sea and the Sulu and Sulawesi seas. However, the piracy-prone areas also include national archipelagic waters where Indonesia will not accept the operational involvement of other countries.

Information exchange between neighbouring countries is another area that requires attention. Some measures are already in place, for example, the ReCAAP Information Sharing Centre, the SURPIC surveillance system in the Singapore Strait and the MSP-IS patrol information system in the Malacca Strait, but these fall well short of the near-real-time coordination of surveillance that would tie in data from all sources, including aerial surveillance, satellite surveillance and shore-based radar systems. The Singapore Maritime Security Centre being established by the Singapore navy, with its Information Fusion Centre, is an example of what may be possible in the longer term (Huang 2008: 98). Designed as a one-stop information and response coordination centre to meet the maritime security needs of Singapore, it is envisioned as a platform for fostering regional information sharing between navies and other agencies. However, sovereign sensitivities remain a barrier to such cooperation in practice. While much has been done in recent years to improve the situation with piracy and armed robbery against ships in Indonesian waters, attacks still occur. There was a resurgence of attacks in 2008 in and around the Riau Archipelago and the southern part of the South China Sea. The situation may deteriorate further, with an Indonesian naval spokesman predicting that the current economic recession will increase the risks to shipping in the Malacca Strait (IMB 2009: 43). Cutbacks in the international shipping industry, for instance, could increase the number of unemployed seafarers in Indonesia and the Philippines, both countries that supply large numbers of seafarers to international shipping. In the longer term, piracy will never be totally eradicated without programs to address the root causes of illegal activity at sea, such as poverty and the lack of economic opportunity in coastal communities.

## ACKNOWLEDGMENTS

The author is indebted to Dr Eric Frécon, post-doctoral fellow in the Indonesia Programme at S. Rajaratnam School of International Studies, Singapore, for his insights into piracy in Indonesia, particularly in and around the Riau Archipelago.

## REFERENCES

Bateman, Sam and Joshua Ho (2008), 'Somalia-type piracy: why it will not happen in Southeast Asia', RSIS Commentary 123/2008, S. Rajaratnam School of International Studies, Singapore, 24 November.

Bateman, Sam, Joshua Ho and Mathew Mathai (2007), 'Shipping patterns in the Malacca and Singapore straits: an assessment of the risks to different types of vessel', *Contemporary Southeast Asia*, 29(2): 309–32.

Bateman, Sam, Catherine Zara Raymond and Joshua Ho (2006), 'Safety and security in the Malacca and Singapore straits: an agenda for action', IDSS Commentary 41/2006, Institute of Defence and Strategic Studies, Singapore, 29 May.

Beckman, Robert C. (2008), 'The 1988 SUA Convention and 2005 SUA Protocol: tools to combat piracy, armed robbery and maritime terrorism', in R. Herbert-Burns, S. Bateman and P. Lehr (eds), *Lloyd's MIU Handbook of Maritime Security*, CRC Press, Boca Raton, pp. 187–200.

Blyth, Captain Ken with Peter Corris (2000), *Petro Pirates: The Hijacking of the Petro Ranger*, Allen & Unwin, Sydney.

Bradford, John F. (2008), 'Shifting the tides against piracy in Southeast Asian Waters', *Asian Survey*, 48(3): 473–91.

Chalk, Peter (2008), *The Maritime Dimension of International Security: Terrorism, Piracy and Challenges for the United States*, Rand Corporation, Santa Monica.

Dirhamsyah, Dirham (2005), 'Maritime law enforcement and compliance in Indonesia: problems and recommendations', Maritime Studies 144, Australian Association for Maritime Affairs, Canberra, September/October.

Eklöf, Stefan (2006), *Pirates in Paradise: A Modern History of Southeast Asia's Maritime Marauders*, NIAS Press, Copenhagen.

Frécon, Eric (2008), '"Pirates of the Riau": the making of an up-and-down story', unpublished paper, S. Rajaratnam School of International Studies, Singapore.

Frécon, Eric (2009), 'Piracy in the South China Sea: maritime ambushes off the Mangkai Passage', RSIS Commentary 18/2009, S. Rajaratnam School of International Studies, Singapore, 20 February.

Hammick, Denise (2007), 'Turning the tide: maritime security in Southeast Asia', *Jane's Defence Weekly*, 28 November.

Ho, Joshua (2006), 'The IMO–KL meeting on the straits of Malacca and Singapore: major maritime nations and stakeholders need to do more', IDSS Commentary 107/2006, Institute of Defence and Strategic Studies, Singapore, 5 October.

Ho, Joshua (2009), 'Piracy in the Gulf of Aden: lessons from the Malacca Strait', RSIS Commentary 18/2009, S. Rajaratnam School of International Studies, Singapore, 22 January.

Huang, Major Victor, RSN (2008), 'Building maritime security in Southeast Asia: outsiders not welcome?', *Naval War College Review*, 61(1): 87–105.

IMB (International Maritime Bureau) (2004), *Piracy and Armed Robbery against Ships: Annual Report for the Period 1 January – 31 December 2003*, International Maritime Bureau, International Chamber of Commerce, London.

IMB (International Maritime Bureau) (2009), *Piracy and Armed Robbery against Ships: Annual Report for the Period 1 January – 31 December 2008*, International Maritime Bureau, International Chamber of Commerce, London.

Jones, Steven M. (2008), 'Implications and effects of maritime security on the operation and management of merchant vessels', in R. Herbert-Burns, S. Bateman and P. Lehr (eds), *Lloyd's MIU Handbook of Maritime Security*, CRC Press, Boca Raton, pp. 87–116.

Lloyd's Joint War Committee (2009), 'Hull war, strikes, terrorism and related perils', Circular JWLA 009, 7 January, available at http://www.lmalloyds.com/AM/Template.cfm?Section=Joint_War1&CONTENTID=15862&TEMPLATE=/CM/ContentDisplay.cfm, accessed 14 January 2009.

Mak, J.N. (2007), 'Pirates, renegades and fishermen: the politics of "sustainable" piracy in the Straits of Malacca', in Peter Lehr (ed.), *Violence at Sea: Piracy in the Age of Global Terrorism*, Routledge, New York, pp. 199–223.

Mak, J.N. (2008), 'Pirates, renegades, and fishermen: reassessing the dynamics of maritime piracy in the Malacca Straits', paper presented to the Royal Australian Navy Sea Power Conference 2008, Sydney, 29–31 January.

Murphy, Martin N. (2007), 'Contemporary piracy and maritime terrorism', Adelphi Paper 388, International Institute of Strategic Studies, London.

Raymond, Catherine Zara (2007), 'Piracy in the waters of Southeast Asia', in K.C. Guan and J.K. Skogan (eds), *Maritime Security in Southeast Asia*, Routledge, Abingdon, pp. 62–77.

Raymond, Catherine Zara and Arthur Morrien (2008), 'Security in the maritime domain and its evolution since 9/11', in R. Herbert-Burns, S. Bateman and P. Lehr (eds), *Lloyd's MIU Handbook of Maritime Security*, CRC Press, Boca Raton, pp. 3–12.

ReCAAP (Regional Cooperation Agreement on Combating Piracy and Armed Robbery against Ships in Asia) (2007), 'Report for January 2007', ReCAAP Information Sharing Centre, Singapore, available at http://www.recaap.org/incident/reports.html.

ReCAAP (Regional Cooperation Agreement on Combating Piracy and Armed Robbery against Ships in Asia) (2008a), 'Report for October 2008', ReCAAP Information Sharing Centre, Singapore, available at http://www.recaap.org/incident/reports.html.

ReCAAP (Regional Cooperation Agreement on Combating Piracy and Armed Robbery against Ships in Asia) (2008b), 'Report for November 2008', ReCAAP Information Sharing Centre, Singapore, available at http://www.recaap.org/incident/reports.html.

ReCAAP (Regional Cooperation Agreement on Combating Piracy and Armed Robbery against Ships in Asia) (2008c), 'Special report on the hijacking of the *Whale 7*', ReCAAP Information Sharing Centre, Singapore, available at http://www.recaap.org/incident/reports.html.

ReCAAP (Regional Cooperation Agreement on Combating Piracy and Armed Robbery against Ships in Asia) (2008d), 'Annual report 2008', ReCAAP Information Sharing Centre, Singapore, available at http://www.recaap.org/incident/reports.html.

Sasakawa, Yohei (2007), 'Towards a new world maritime community: cooperative framework for the future of the Malacca and Singapore straits', RSIS Commentary 17/2007, S. Rajaratnam School of International Studies, Singapore, 21 March.

Storey, Ian (2008), 'Securing Southeast Asia's sea lanes: a work in progress', *Asia Policy*, 6(July): 95–127.

Timlen, Thomas (2008), 'The use of SOLAS ship security alert systems', RSIS Working Paper No. 154, S. Rajaratnam School of International Studies, Singapore, 5 March.

Tsamenyi, Martin, Mary Ann Palma and Clive Schofield (2008), 'International legal regulatory framework for seafarers and maritime security post 9/11' in R. Herbert-Burns, S. Bateman and P. Lehr (eds), *Lloyd's MIU Handbook of Maritime Security*, CRC Press, Boca Raton, pp. 233–52.

# 8 THE INDONESIAN MARITIME SECURITY COORDINATING BOARD

*Djoko Sumaryono*

---

Indonesia's archipelagic character means that its seas are a strategic resource of immense importance. The seas function as a means to unite the nation, as a means of transportation, as a means of defence and security, and as a means of diplomacy. They are also a source of important natural resources that help to sustain the nation's economic well-being. Guarding the seas is therefore an important part of guarding Indonesia's state sovereignty. Indonesia has the sovereign right to regulate, control, protect and manage its sea territory in order to safeguard its national interest. But it cannot do this unless it has in place a comprehensive and integrated maritime security system. This is the setting in which Indonesia established, and then injected new life into, the Maritime Security Coordinating Board.

## BACKGROUND

Indonesia is situated at a strategic crossroads (see Map 1.1 on page 2 and Map 2.1 on page 29). About 40 per cent of the world's sea trade is carried on its sea lanes. Of particular importance is the high volume of shipments of crude oil transported through Indonesian seas. The total value of trade passing through Indonesian archipelagic sea lanes amounts to US$300 trillion annually in the Malacca Strait, US$40 trillion annually in the Lombok Strait and US$5 trillion annually in the Sunda Strait. Indonesia's archipelago is also of great importance to the world's climate. Streams from the Pacific and Indian Ocean meet in the archipelago, where they not only create a rich natural ecosystem but also, because of the mixing and circulation of waters, have an important though still unquantified effect on global climate change.

Responsibility for maritime affairs is currently spread across a number of government institutions: Customs is under the authority of the Directorate General of Customs and Excise in the Ministry of Finance; immigration is the domain of the Directorate General of Immigration in the Ministry of Justice and Human Rights; fisheries are the under the authority of the Ministry of Marine Affairs and Fisheries; mangrove forests are the responsibility of the Ministry of Forestry; and so on.

Under these circumstances, it is not easy to avoid overlaps of authority. It sometimes happens that inquiries into maritime cases are conducted by officers from the navy, the marine police and the attorney-general's office as well as assigned officers from technical departments. This makes it very difficult to coordinate policies and plans that cut across all jurisdictions or to implement programs to control marine resources. There has been a tendency for officers assigned to the field to give higher priority to the programs and policies of their own institutions than to those of other institutions.

In the early years of the New Order, it was apparent that Indonesia needed a body to coordinate the handling of maritime security issues. This led to the founding in 1972 of the Maritime Security Coordinating Board (Badan Koordinasi Keamanan Laut, or Bakorkamla), created by a joint ministerial decree of the ministers of defence, communications, finance and justice, as well as the attorney-general.[1] The new body was chaired by the defence minister. In addition to the ministers of communications, finance and justice and the attorney-general, it included representatives from navy headquarters, police headquarters and the Directorate-General of Fisheries in the Ministry of Agriculture.

Unfortunately this body did not live up to its potential, and many aspects of marine security policy remained uncoordinated. The first Maritime Security Coordinating Board tended to focus on issues of maritime security, and generally did not address economic and marine environmental protection issues. In October 2002, the then Coordinating Minister for Political and Security Affairs, Susilo Bambang Yudhoyono, commented bluntly that 'In terms of weaponry, instruments, technological equipment, detection and pursuit [of trespassers and criminals], we are weak' (Antara, 9 October 2002). In 2003, the minister formed a working group to find ways to improve security and law enforcement at sea. Over the ensuing two years, this group conducted a series of seminars and consulted widely across government departments to lay the foundation for a more effective agency to coordinate maritime security. On

---

1    See http://www.bakorkamla.go.id/umum_eng.php.

29 December 2005, Presidential Regulation No. 81/2005 on the Maritime Security Coordinating Board established the legal basis for the present organization.[2]

## STRUCTURE

The structure of the revitalized Maritime Security Coordinating Board reflects the necessity to protect and safeguard Indonesia's national integrity and marine assets, which are increasingly threatened by illegal activity at sea. The reinvigoration of the Maritime Security Coordinating Board is driven not only by domestic needs but also by Indonesia's developing international responsibilities. The Indonesian seas include some of the world's most important and heavily used maritime trade routes. As a longstanding member of the International Maritime Organization (IMO) and as a signatory to the 1982 United Nations Convention on the Law of the Sea (UNCLOS), Indonesia has an obligation to ensure the safety and security of the vessels using its waters by setting up the appropriate national and local authorities (UNCLOS, articles 217, 218 and 220).

The new Maritime Security Coordinating Board is directly responsible to the president. It has the somewhat daunting task of coordinating the 12 ministerial-level agencies that share responsibility for maritime safety and security. Board members meet regularly each month at the organization's headquarters to discuss any incidents or developments that have taken place within their jurisdictions during the preceding weeks, and the measures they have taken in response. Given the perceived importance of the organization's work, ministers and agency heads usually attend these meetings, although occasionally they may be represented by officials at the director-general level. The Maritime Security Coordinating Board also holds regular forums on maritime security operations to which maritime experts are invited. This practice reflects the agency's interest in looking beyond departmental confines to consult widely with stakeholders in a constructive and cooperative way.

The organizational structure of the Maritime Security Coordinating Board is set out in Figure 8.1. It is chaired by the Coordinating Minister for Political, Legal and Security Affairs. The 12 member ministries and agencies are: (1) the Coordinating Ministry for Political, Legal and Security Affairs; (2) the Ministry of Foreign Affairs; (3) the Ministry of the Interior; (4) the Ministry of Defence; (5) the Ministry of Justice and Human Rights; (6) the Ministry of Finance; (7) the Ministry of Transport; (8) the Ministry of Marine Affairs and Fisheries; (9) the Office of the

---

2   See www.bakorkamla.go.id/doc/PerpresNo81th2005.pdf.

Figure 8.1    *The Maritime Security Coordinating Board: organizational chart*

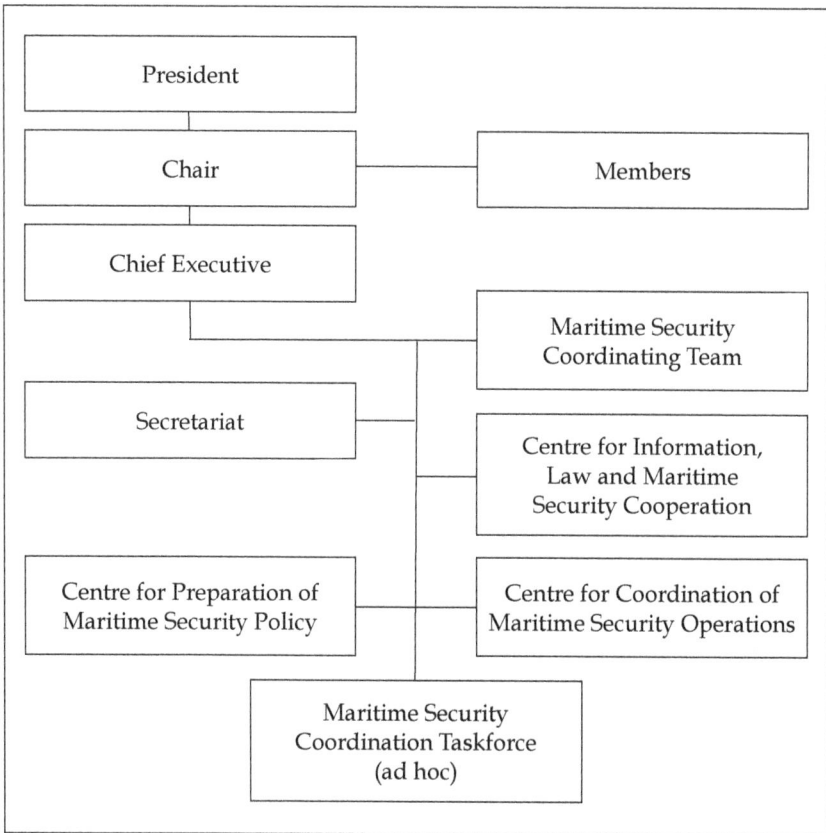

Attorney-General; (10) the Armed Forces; (11) the National Police; and (12) the State Intelligence Agency. The six implementing agencies are: (1) the Maritime Security Coordinating Team (represented by the upper echelon of each member ministry or agency); (2) the Secretariat of the Maritime Security Coordinating Board; (3) the Centre for Preparation of Maritime Security Policy; (4) the Centre for Information, Law and Maritime Security Cooperation; (5) the Centre for Coordination of Maritime Security Operations; and (6) the Maritime Security Coordination Task Force (ad hoc).

## GOALS

The overarching goal or vision of the Maritime Security Coordinating Board is to create an integrated system of maritime security, safety and law enforcement for the areas within Indonesia's maritime jurisdiction. Within this framework, the organization has been given five specific tasks:

1   to formulate general policy on maritime security;
2   to coordinate maritime security operations in Indonesian waters;
3   to provide technical and administrative support for maritime security;
4   to provide assistance in maritime security institutional capacity building; and
5   to encourage stakeholder engagement in ensuring maritime security.

The Maritime Security Coordinating Board is not directly involved in law enforcement action, but it does play a role in coordinating the operations of member institutions. It coordinates monitoring, surveillance, protection and law enforcement for these operations, to secure the safety of vessels and passengers travelling through Indonesian waters.

The revitalization of the Maritime Security Coordinating Board in 2005 has sharpened its organizational structure, increased its capacity to carry out its tasks and put in place clearer arrangements and operational mechanisms for members. The strengthening of the legal foundations of the organization has made it easier for members to coordinate and integrate their activities. This has allowed the new body to become more effective not only in pursuing security-related matters, but also in reducing the economic losses caused by illegal activities at sea and in strengthening marine environmental protection.

On 20 July 2007, the Maritime Security Coordinating Board adopted a strategic plan for the next two years. The plan sets out the values of the organization, its strategic goals and the way in which resources will be allocated, backed by a set of performance indicators. The purpose of having a clear set of organizational values is to build a work culture that supports the achievement of the agency's goals. These values are built around the principles of rule of law, good governance, public interest, openness, proportionality, professionalism and accountability. In conducting its duties and functions, the Maritime Security Coordinating Board has committed itself to reforming and improving the maritime security sector.

## STRENGTHENING MARITIME SECURITY

Maritime security issues are the main focus of the Maritime Security Coordinating Board. These include smuggling, sea piracy, human trafficking, terrorism and illegal fishing. It is estimated that around half a million small arms and light weapons are smuggled through Southeast Asia each year, with 80 per cent of the activity carried out by sea. Although the frequency of piracy and armed robbery at sea has decreased since 2003, the International Maritime Bureau reported 43 actual and attempted attacks in Indonesian waters in 2007, seven of them in the Malacca Strait alone. Most of these attacks occurred while the vessels were in port or at anchor.

Illegal fishing in Indonesian waters has a huge impact on the domestic economy, causing losses estimated at around US$4–5 billion annually. Some 3,180 foreign fishing vessels are thought to be operating illegally in Indonesian waters. Illegal logging causes additional losses of around US$700 million annually. The losses in 2007 resulting from illegal fishing, mining and logging together have been estimated at Rp 617 trillion (US$670 million). Three operations conducted by the Maritime Security Coordinating Board in 2007 are thought to have reduced these losses by an estimated Rp 118.65 billion (US$13 million).

In 2007 and 2008, the Maritime Security Coordinating Board conducted four operations in Indonesian seas to protect legitimate domestic economic activity and take action against those who violate the country's laws. In addition to the obvious benefit of reducing the potential losses from illegal activities, these operations had a number of positive outcomes. First, they represented a significant strengthening of maritime law and order. This is not only good for business but helps to build trust among the people. Both are essential preconditions for ensuring the sustainability of natural resources, marine environment protection and continuing development. Second, at the institutional level, the operations coordinated by the Maritime Security Coordinating Board proved valuable in strengthening the capacity of the relevant institutions to carry out complex law enforcement operations at sea in a harmonious and effective manner. The lessons learned from them are being used to formulate and refine general maritime security policy. Finally, the coordinated operations showed all sea users that the Indonesian government is prepared to enforce the law at sea, regardless of the difficulties.

The four operations coordinated by the Maritime Security Coordinating Board — codenamed Octopus 1, 2, 3 and 4 — are described briefly below. In carrying out these operations, the agency paid particular attention to the need to prevent information about its activities from leaking in advance. In the past, the illegal or careless release of information by one or other of the institutions involved had given maritime criminals sufficient warning to allow them to flee before an operation took place. To prevent this from happening again, the Maritime Security Coordinating Board worked with the State Code Institute (Lembaga Sandi Negara) to develop ways to secure information from leaks, including conducting intelligence activities before each operation got under way.

## Octopus 1 (April–May 2007)

Octopus 1 was carried out over 20 working days in April and May 2007. The parties involved in the operation were the Maritime Security Coordinating Board, the navy, the marine police, Customs and the Ministry of Marine Affairs and Fisheries. Over the three weeks of the operation, crew members of nine of the 59 vessels detained were charged with offences. The potential losses averted by this operation are estimated at Rp 10 billion (US$1.1 million) in tobacco excise, Rp 50 billion (US$5.5 million) in fish and Rp 0.58 billion (US$63,000) in tin, to which can be added the money generated by auctioning some of the ships involved, namely Rp 5 billion (US$544.000). This brings the total to Rp 65.58 billion ($7.2 million).

## Octopus 2 (June–July 2007)

Octopus 2 was carried out over 20 working days in June and July 2007. It focused on the Seram, Halmahera, Sulawesi and Arafuru seas, given that illegal logging and illegal fishing are known to be widespread in the Maluku area. This time, a greater number of Maritime Security Coordinating Board personnel were involved, together with personnel from other member institutions, most of them investigating officers. The operation was supported by Australian Border Protection Command, which provided information from its satellite monitoring system on ships operating in the maritime border zone between the two countries. Ships and helicopters from the navy, the marine police, the Ministry of Transport and Customs were also mobilized for this operation. Despite rough seas, the team performed well, detaining 151 vessels suspected of illegal activities. The potential losses averted through this operation have been estimated at just over Rp 60 billion: Rp 59.60 billion from illegal fishing, Rp 0.40 billion from smuggled goods and Rp 0.12 billion from illegal logging.

## Octopus 3 (October–November 2007)

Octopus 3 was conducted from 30 October to 18 November 2007 in the Malacca Strait, the Natuna Sea and the Bangka Strait. It involved personnel from the Ministry of Marine Affairs and Fisheries, Customs, the marine police, the army and the navy. It is estimated that their actions may have saved the country around Rp 10.6 billion per day. The officers involved in this operation encountered conspicuous violence as they tackled goods smuggling off Tanjung Balai Karimun, the transport of illegal immigrants, the use of illegal or forged oil and gas permits and acts of marine environment pollution.

## Octopus 4 (March–April 2008)

The Octopus 4 operation was conducted in the seas of eastern Indonesia from 12 March to 2 April 2008. The focus was on violations of law at sea, in particular illegal fishing and the carriage of illegally obtained logs, which are known to take place in the vicinity of the Aru Islands, the Tanimbar Islands, Tual in the Kei Islands, and Timika and Merauke in Papua. The operation again used vessels and aircraft mobilized by the navy, the Ministry of Transport and the Ministry of Marine Affairs and Fisheries to patrol the Arafuru, Molucca, Halmahera and Sulawesi seas. Personnel from the State Code Institute were involved in this operation, which resulted in 72 vessels being inspected and 12 of them being detained. The revenue losses averted by this operation are estimated at Rp 335.2 billion.

## Other operations

In addition to the four Octopus operations, the Maritime Security Coordinating Board has conducted a number of operations to support the regular activities of Customs and the Ministry of Marine Affairs and Fisheries. These were conducted off Surabaya and Maluku and in the Seram and Halmahera seas.

## INFORMATION GATHERING AND COOPERATION WITH PARTNERS

As part of its brief to strengthen maritime security and ensure the safety of ships at sea, the Maritime Security Coordinating Board conducts research into shipping accidents, coordinates investigations and cooperates with national and international agencies.

The agency's Centre for Preparation of Maritime Security Policy has just completed a study of the major maritime accidents that have occurred in Indonesia in the past few years. Two major accidents occurred in early 2007, involving the ferries KMP *Senopati Nusantara* and KM *Levina I*. In all, 500 lives were lost. In the case of the KM *Levina I*, analyses showed that the safety criteria for Indonesian ferry services were deficient at both the regulatory and enforcement levels. Roll-on/roll-off ferries were being used to carry far more cargo and passengers than the levels for which they had originally been designed and built. These violations of safety standards were accompanied by other kinds of violations, such as the failure to provide or maintain fire-fighting equipment and life-saving appliances, inadequate cargo-stowing arrangements, poor crew competency and so on. The recommendations arising from this investigation will be used to assist the National Transport Safety Committee (Komite Nasional Keselamatan Transportasi, or KNKT) to draw up more effective regulations covering vessels at sea and tighter standards to ensure that those regulations are adhered to.

To guide the handling of investigations into maritime security incidents, the Maritime Security Coordinating Board has adopted a set of information-handling procedures (see Figure 8.2). To better coordinate its work with that of its members, it is also developing a system to integrate the information it collects. The Integrated Information System has two main functions. The first is to use satellite imagery to track domestic and foreign vessels passing through Indonesian waters, both to protect ships at sea and to support effective and efficient intelligence operations. The second is to gather and process information so that it can be shared with Maritime Security Coordinating Board members and, to the extent possible, with the public. The Integrated Information System is also being used to support search and rescue activities, monitor maritime traffic, protect fisheries, broadcast marine safety information and monitor marine pollution.

As a relatively new organization, the Maritime Security Coordinating Board has actively sought to conduct cooperative programs with overseas and domestic counterparts. To raise international awareness of its activities, the organization's chief executive and other high-level officials have visited Australia, China, Malaysia, Switzerland, Norway, Japan, the Philippines and Germany, as well as hosting visits from officials from Australia, Japan, China, the Russian Federation and the United States, among others. Staff have also attended various international meetings, including the Asian Coast Guard Agencies Meeting in Singapore in November 2007, which was attended by representatives from 18 countries.

Indonesia wants other nations to be aware that it is committed to executing its duties and responsibilities according to international maritime conventions on safety and security. To this end, the Maritime Security Coordinating Board presented a report on the structure and activities of the revitalized agency to the 25th Assembly of the IMO in November 2007.

The Maritime Security Coordinating Board's ability to liaise with other countries' agencies during a crisis was tested when the Panama-

Figure 8.2    *The Maritime Security Coordinating Board: information handling system*

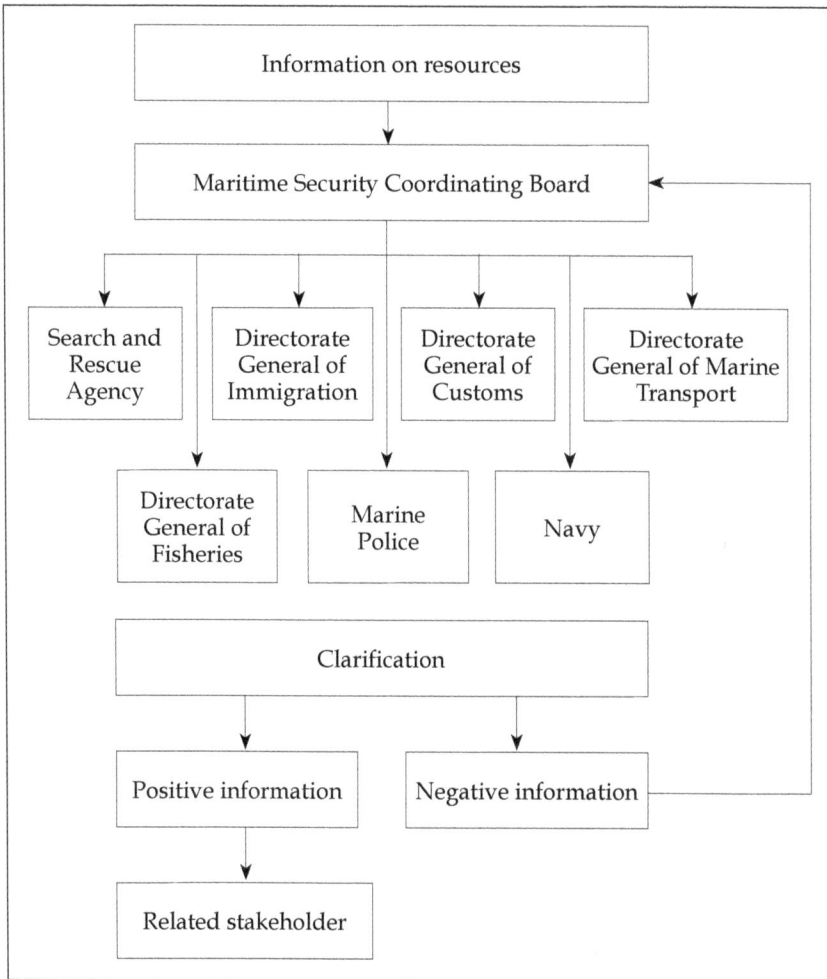

```
┌─────────────────────────────────────────────────────────────────────────┐
│                                                                           │
│              ┌───────────────────────────────────────┐                    │
│              │        Information on resources        │                    │
│              └───────────────────┬───────────────────┘                    │
│                                  ▼                                         │
│         ┌──────────────────────────────────────────┐                      │
│         │   Maritime Security Coordinating Board    │◄──────┐              │
│         └──┬──────────┬───────────────┬──────────┬──┘       │              │
│            ▼          ▼               ▼          ▼          │              │
│     ┌──────────┐ ┌──────────┐ ┌──────────┐ ┌──────────┐    │              │
│     │Search and│ │Directorate│ │Directorate│ │Directorate│  │             │
│     │ Rescue   │ │General of│ │General of│ │General of │    │             │
│     │ Agency   │ │Immigration│ │ Customs │ │Marine    │     │             │
│     └──────────┘ └──────────┘ └──────────┘ │Transport │     │             │
│          ▼             ▼            ▼        └──────────┘    │             │
│     ┌──────────┐ ┌──────────┐ ┌──────────┐                  │             │
│     │Directorate│ │  Marine  │ │  Navy   │                  │             │
│     │General of │ │  Police  │ │         │                  │             │
│     │ Fisheries │ │          │ │         │                  │             │
│     └──────────┘ └──────────┘ └──────────┘                  │             │
│         ┌──────────────────────────────────────┐            │             │
│         │            Clarification              │            │             │
│         └──────┬───────────────────┬───────────┘            │             │
│                ▼                   ▼                         │             │
│     ┌──────────────────┐ ┌──────────────────┐               │             │
│     │Positive          │ │Negative          │───────────────┘             │
│     │information       │ │information       │                             │
│     └────────┬─────────┘ └──────────────────┘                             │
│              ▼                                                             │
│     ┌──────────────────┐                                                  │
│     │Related stakeholder│                                                 │
│     └──────────────────┘                                                  │
└─────────────────────────────────────────────────────────────────────────┘
```

*Explanation:*

1   Upon being advised of a maritime incident, the Maritime Security Coordinating Board will instruct the relevant government ministry or agency to investigate the incident.

2   The relevant ministry or government agency will report back to the Maritime Security Coordinating Board on its findings.

3   Upon receiving such a report, the Maritime Security Coordinating Board will independently clarify and confirm its findings and make recommendations if necessary.

4   If further action is needed, the relevant ministry or agency will carry this out and report its findings to the Maritime Security Coordinating Board.

5   If no further action is needed, the Maritime Security Coordinating Board will so advise the relevant ministry or agency.

6   The Maritime Security Coordinating Board will ensure that ministerial and other agencies are kept informed of the way in which information is being handled.

registered freighter *Mezzanine* encountered rough seas north of Taiwan in November 2007. The agency communicated intensively with Taiwan's coast guard to secure the rescue of the Indonesian crew.

To increase awareness of its activities among Indonesians, in December 2007 the Maritime Security Coordinating Board held a seminar in Bogor on its efforts to enhance maritime security in Indonesian waters. The seminar was attended by domestic stakeholders throughout Indonesia.

The Maritime Security Coordinating Board regards human resources as crucial for the successful implementation and discharge of its duties. Together with the Ministry of State Apparatus and the Ministry of Education, it is studying the feasibility of establishing a coast-guard academy to support the anticipated establishment of a dedicated coast guard. Japan has agreed to support this plan through the Japan International Cooperation Agency.

## ESTABLISHING A SEA AND COAST GUARD

On 8 April 2008, the Indonesian parliament passed Law No. 17/2008 on Shipping. It provided for the establishment of a Sea and Coast Guard (Kesatuan Penjaga Laut dan Pantai, or KPLP) with responsibility for ensuring maritime safety and security in the maritime areas that lie within Indonesia's jurisdiction. The duties of the Sea and Coast Guard will be:

- to conduct oversight to ensure maritime safety and security;
- to conduct oversight to prevent and combat marine pollution;
- to monitor and control maritime traffic;
- to monitor and control salvage activity, underwater work and exploration and exploitation activities;
- to protect navigational aids; and
- to support maritime search and rescue activities.

The Maritime Security Coordinating Board, meanwhile, has broad oversight of sea security and relevant policy.

The work of the Maritime Security Coordinating Board described above provides the context for the establishment of the Sea and Coast Guard. The agency has performed well during its first few years of operation and has achieved much for the country. However, as the threats to maritime security grow, so does the need for a single body with more effective authority. The new Sea and Coast Guard will need to be characterized by a single command with a defined range of functions and the capacity to marshal the resources and skills of the institutions that have control and oversight powers over activities at sea.

The Maritime Security Coordinating Board has begun drafting a government regulation to provide the legal basis for the establishment of a Sea and Coast Guard. Together, the Sea and Coast Guard and the reinvigorated Maritime Security Coordinating Board will strengthen the framework for ensuring maritime safety and security in Indonesian waters. Indonesia is committed to ensuring the safety and security of its seas and to working hand in hand with other countries who share the same goal.

# 9 MARINE SAFETY IN INDONESIAN WATERS

*Erwin Rosmali*

Managing the safety of those who go to sea in Indonesian waters is an important task of the Indonesian government. In the past, Indonesia did not enjoy a good reputation in this field, but today the government is making an increased effort to uphold international marine safety standards.

Seafarers and travellers in Indonesian waters face a complex range of hazards. Because Indonesia is an archipelago, many of its sea lanes run through narrow passages between islands, with consequent problems of congestion (see Map 1.1 on page 2). This issue is particularly serious in the Malacca Strait, which Indonesia shares with Malaysia. The strait is 1,000 kilometres long and, at some points, only two kilometres wide and 23 metres deep. It is the main line of sea communication between the Indian Ocean and the South China Sea. Of the oil transported from the Middle East to East Asia, 85 per cent passes through this narrow corridor. Each year about 60,000 ships, with an estimated weight of 4 billion deadweight tonnes, sail through the strait. Another 14,000 ships, including supertankers, pass through the Lombok Strait, which is both broader (at 11 kilometres) and deeper than the Malacca Strait. A further 3,500 ships per year use the Sunda Strait (Ho 2006: 559–61). Although much shipping follows major routes, there is a significant volume of traffic on smaller routes, many of which cross the larger ones, creating further congestion problems.

According to the latest figures from the Directorate of Marine Traffic in the Directorate General of Marine Transport (31 March 2008), Indonesia's national fleet comprises 7,846 ships. This is a 29.9 per cent increase over the previous total of 6,041 ships, recorded in March 2005. However, the statistic is misleading, because it reflects the registration of many ships which had previously operated under flags of convenience in order to evade the Indonesian government's requirements on safety and labour (Sijabat 2007b). In 2005, the government required all Indonesian vessels operating in the nation's waters to operate under the national flag, forcing vessels flying flags of convenience, newly built ships and used ships procured from abroad to register as Indonesian vessels.[1]

Although most of Indonesia lies outside the tropical cyclone zone, shipping plying Indonesian waters faces the problem of sudden equatorial storms. These are typically a late afternoon phenomenon, the result of heating of air and of consequent convection currents which lead to short-lived but strong local winds. The exact location of such storms is difficult to predict and shipping can easily be caught unawares. The problem is exacerbated by the abundance of coral reefs and shallow banks in Indonesian seas, which can be both a direct hazard and, in bad weather, a cause of dangerous waves. Although Indonesia has an extensive network of lighthouses, they fall far short of covering all the natural hazards that ideally would be marked in the archipelago.

These hazards have been made more serious by a lack of investment in measures that would improve marine safety. Few ships have up-to-date navigational equipment and many do not carry a full complement of safety equipment. In addition, crews on some larger ships have had relatively little training in handling emergencies, and shipowners are sometimes negligent in maintaining the seaworthiness of their vessels. It was recently reported that half the ferries operating on the busy route across the Sunda Strait between Java and Sumatra were built as long ago as the 1970s or 1980s (Fidrus 2008). Ferries are often significantly overloaded, making the death toll in disasters greater than it would be if loading limits were adhered to.

The result is a tragic catalogue of major maritime disasters. In modern times, the major ones are as follows (in order of number of lives lost):

1981 (January): A 25-year-old ferry, the *Tampomas II*, caught fire and sank near Masalembo Island in the Java Sea, resulting in the loss of 580 lives.

2000 (June): The *Cahaya Bahari*, which was carrying refugees from communal conflict in northern Maluku, sank about 80 kilometres north of Manado with the loss of 550 lives.

---

1   According to figures provided by the United Nations Conference on Trade and Development (UNCTAD), the total tonnage of Indonesia-flagged ships was 5,287,148 gross tonnes in 2008, giving Indonesia the twenty-fourth largest fleet in the world.

2006 (December): The car ferry *Senopati Nusantara* sank off Mandalika Island in the Java Sea with the loss of 461 lives.

1999 (October): The *Bismas Raya II* caught fire and sank off Merauke, Papua, with the loss of 361 lives.

2001 (October): An Indonesian fishing vessel—later called the *SIEV-X* (Suspected Illegal Entry Vessel No. 10)—sank in bad weather about 51.5 nautical miles south of Java with the loss of 353 lives. The vessel had departed from Bandar Lampung and was carrying asylum seekers bound for Australia. The disaster occurred in international waters, but the site lay within Indonesia's EEZ and was within Indonesia's internationally designated zone of search and rescue responsibility.

1996 (January): A passenger ferry, the *Gurita*, sank off Sabang, Aceh, causing 338 deaths.

1999 (February): An unlicensed passenger ferry, the *Harta Rimba*, capsized off the coast of West Kalimantan in the South China Sea, with at least 325 people confirmed dead.

2009 (January): A passenger ferry, the *Teratai Prima*, capsized off Sulawesi, causing about 230 deaths.

2005 (July): About 200 people died when a passenger ferry, the *Digul*, capsized in rough seas off Merauke, Papua.

2002 (November): The *Masori Star* capsized off Ambon, with at least 77 people confirmed dead.

2007 (October): The *Acita III* capsized off Baubau, Sulawesi, with at least 66 people confirmed dead.

2006 (June): The *Surya Makumur Indah* capsized off Sibolga, North Sumatra, with at least 98 people rescued and 35 confirmed dead.

In addition to these major disasters, there have been dozens of incidents in which boats carrying small crews have been lost at sea. Until recently, Indonesia's response to such disasters was slow and piecemeal. The high cost of maintaining a safety regime at sea has been a serious obstacle to providing sound support for marine safety in Indonesia. Corruption in the administration of marine safety has also been a problem (Soedarjo 2007).

Historically, governments have used a variety of techniques to improve safety at sea.[2] These include measures to ensure that ships are basically seaworthy and that skippers and crew are qualified, regulations

---

2  On the history of marine safety regulation, see Boisson (1999).

governing behaviour (such as load limits, the requirement to carry safety equipment and navigational lights, the rule of passing to the right at sea, rules of rescue and so on) and the provision of navigational aids such as lighthouses and, more recently, various forms of radio communication. Indonesia became a member of the International Civil Aviation Organization in 1950 and joined the International Maritime Organization (IMO) in 1961. In recognition of the country's obligations as a member of these organizations to respond to air and sea accidents, in 1968 the Minister of Transport ordered the formation of a small search and rescue team in Jakarta. This measure was followed in 1972 by the creation of the Indonesian Search and Rescue Agency (Badan SAR Indonesia, or Basari), which, however, focused primarily on air disasters.

In the late 1970s, Indonesia risked being declared a 'black area' for international transport because of its poor search and rescue facilities. In 1979, the government responded by forming a new body, the National Search and Rescue Agency (Badan SAR Nasional, or Basarnas), within the Ministry of Transport.[3] The sinking of the *Tampomas II* in 1981 prompted an immediate ban on the use of roll-on/roll-off ferries (Rutz and Coull 1996: 279). In 1999, the government founded a National Transport Safety Committee (Komite Nasional Keselamatan Transportasi, or KNKT) within the Ministry of Transport, but its mandate initially did not include maritime incidents. Between 2004 and 2006, the committee had the resources to investigate only one of the 24 marine accidents that had taken place during that period, although it investigated 22 of 26 rail accidents (Taufiqurrahman 2007). Indonesia has around 180 coastal radio stations broadcasting to ships, but few are on the air 24 hours a day.[4] As a coastal state, Indonesia has the responsibility to provide navigational aids to assist ships, and to conduct search and rescue operations in response to accidents in Indonesian waters. Indonesia needs 4,959 navigational aids — lighthouses, beacons, buoys and so on — but has only 2,780, or just 58.6 per cent of what is required. Of these, 93 per cent are in good order.

In recent years, there have been promising signs of a more vigorous application of marine safety regulations (Sijabat 2007a). As a long-term member of the IMO, Indonesia has ratified various international maritime conventions on marine safety and marine environment protection. Indeed, with so much to gain from better safety at sea, Indonesia has ratified more IMO conventions than any other nation in Southeast Asia.

---

3 'Sejarah SAR Nasional' [A national history of search and rescue], available at http://basarnas.go.id/index.php?page=01, accessed 3 December 2008.
4 See http://www.indomarinav.com/navice/stp_main_stpGrid.php, accessed 2 December 2008

Indonesia's ratification of IMO conventions gives it rights and responsibilities as a flag state, a port state and a coastal state. As a flag state, it has the responsibility to ensure that ships flying the national flag comply with international regulations on technical management and labour. As a port state, Indonesia has the right to investigate the documents and equipment of foreign-flagged ships that use its ports, and the responsibility to provide waste reception facilities for foreign ships in its international ports. At the moment, there are 2,131 designated ports in Indonesia, consisting of 975 regular ports and 1,156 special ports. The four main ports are Belawan (Medan), Tanjung Priok (Jakarta), Tanjung Perak (Surabaya) and Makassar.

In order to improve services and comply with the requirements of international conventions concerning marine safety and security, on 1 July 2004 Indonesia formally applied safety and security standards based on the International Ship and Port Facilities Security (ISPS) Code. At present 677 ships, 237 ports and 16 recognized security organizations have been certified as complying with that code. In 2006, reflecting the growing importance placed on marine safety and security by the national government, Basarnas was taken out of the Ministry of Transport and placed directly under the president.

In June 2007, the Hydro-Oceanography Division (Jawatan Hidro-Oseangrafi, or Janhidros) of the Indonesian navy and the Maritime and Port Authority of Singapore jointly launched an electronic navigational chart covering the seas between Batam and Singapore. From July 2008, each of the 60 or so high-speed ferries operating on the Batam–Singapore route was required to purchase a device enabling it to log on to a series of eight electronic navigational beacons. The head of Janhidros, Commodore Willem Rampangile, has announced plans to extend the system to 18 other ports.[5]

The most important recent development in marine safety has been the passing of Law No. 17/2008 on Shipping on 7 May 2008, designed to replace Law No. 21/1992 on Shipping, which was widely regarded as out of date. It covers four major areas: water transport, ports, marine safety and security, and protection of the marine environment.

The new law will enable Indonesia to comply more effectively with its obligations under the ISPS Code. The code obliges Indonesia to anticipate technological developments in safety and security and refer to international best practice on the use of sophisticated equipment.

---

5    'RI, S'pore launch electronic navigational chart system', *Jakarta Post*, 14 June 2007, available at http://www.thejakartapost.com/news/2007/06/14/ri-s039pore-launch-electronic-navigational-chart-system.html, accessed 1 December 2008.

The most important institutional change to arise from the Shipping Law is the establishment of a coast guard. In the past, some coast-guard functions were carried out by the navy and others by coast guards attached to provincial police offices. Both the navy and the police tended to give low priority to coast-guard work. In West Kalimantan, for instance, coast-guard vessels were able to travel at only half the speed of the illegal fishing boats they were intended to pursue.[6] With the separation of the national police force from the Indonesian army in 1999, the idea of establishing a separate coast guard emerged. The argument for a separate institution strengthened in 2005 when Malaysia established its own coast guard, the Malaysian Maritime Enforcement Agency. The issue was discussed extensively during parliamentary debate on the Shipping Law, with opponents arguing that coast-guard functions were best carried out by the navy and that increased naval spending was a higher priority (Nurhayati 2007). In addition to its command function of maintaining maritime safety and security, the coast guard is charged with coordinating activities to uphold the law in other areas as well. Considerable work remains to be done, however, to determine the proper relationship between the coast guard and other law enforcement bodies. This and other matters will need to be clarified through presidential decrees and other regulations.

## RESPONSIBILITY FOR MARINE SAFETY

The regulation of marine safety in Indonesia is the responsibility of the Directorate of Marine Safety, a branch of the Directorate General of Marine Transport in the Ministry of Transport. The directorate general was established under Ministerial Decree No. 43/2005 of the Minister of Transport. In recognition of the critical importance of shipping to the economic and social health of Indonesia, it aims both to promote the expansion of shipping to serve as many regions as possible within the Indonesian archipelago and to improve the standard of marine safety. Under the new Shipping Law, it is tasked with formulating and implementing policy on ships' seaworthiness, vessel tonnage measurement, ship registration, nautical, technical and radio regulations, pollution prevention, safety management and seafarers' affairs.

One of the marine safety issues currently demanding the close attention of the Directorate of Marine Safety is the process for investigating

---

6  'Vintage coast guard outwits foreign armada', *Jakarta Post*, 10 September 1999, available at http://www.thejakartapost.com/news/1999/09/10/vintage-coast-guard-outwits-foreign-armada.html, accessed 3 December 2008.

shipping accidents, so that effective preventive measures can be taken. It is also attempting to identify and mark currently unmarked sea lanes, and survey those waters still categorized as 'grey' or undersurveyed. The directorate monitors both the volume of traffic and the dimensions of shipping in sea lanes, and seeks to harmonize overlapping and inconsistent regulations.

In seeking to improve Indonesia's capacity to uphold international standards of marine safety, the Directorate of Marine Safety has focused particular attention on the seaworthiness of vessels. Current regulations need to be updated in the light of changing technology and more rigorous international standards. Even more importantly, existing regulations need to be implemented more vigorously and consistently. In particular, the process of inspecting ships for compliance with marine safety regulations needs to be enhanced. In carrying out this task, the directorate must call on the cooperation of many other sections of government, both within and outside the Directorate General of Marine Transport.

This collaboration has been delivering some promising results, with Indonesia making significant progress on enhancing marine safety in several areas.

## Navigation

The Directorate of Navigation is working on the planning, procurement, operations, maintenance and supervision of navigational aids and marine telecommunications based on international regulations. The provision of these facilities is especially important in the archipelagic sea lanes that Indonesia is obliged to recognize under the 1982 United Nations Convention on the Law of the Sea (UNCLOS); Indonesia ratified the convention in 1985 under Law No. 17/1985. In archipelagic sea lanes, the relevant archipelagic state is obliged to provide access to foreign-flagged ships, including suitable navigational aids stationed outside those ships, and designed and operated to increase the safety and efficiency of ship navigation and/or ship traffic. Marine telecommunications includes the broadcasting, sending and receiving of locations, pictures, sound and other information through fibre optic cable, radio or marine safety systems.

## Climate/weather

The Meteorology and Geophysics Agency (Badan Meteorologi dan Geofisika, or BMG) is a non-departmental government body with the role of managing government responsibilities in meteorology, climatology, air quality and geophysics. BMG collects, analyses and distributes data on weather conditions and air quality. In the field of marine safety, one of its most important tasks is to provide wave information.

## Nautical publications

To support marine safety, the government issues nautical publications such as sea charts, guides for pilots, a list of Indonesian lighthouses and lights at sea, a nautical almanac, tide tables and notices to mariners. These are distributed by the Indonesian navy through the Hydrography Institute.

## Search and rescue

With its new status as a non-departmental government body, Basarnas is in charge of handling the response to transport accidents and natural disasters. Hampered by a lack of infrastructure, information and communication tools, and resources, Basarnas must still rely on the cooperation of the army, the police and other bodies. Indonesia's complex geography means that bringing the country's search and rescue service up to international standards will still take some time.

## Sounding and dredging

Many of the marine routes leading into Indonesian ports run through bays, estuaries and rivers that are susceptible to sedimentation. To maintain these routes, depth measurement and dredging are performed regularly.

## Piloting and towing

In order to maintain the safety of navigation in harbours, it is compulsory for vessels to use piloting and towing services, carried out in accordance with the applicable regulations. Pilots must hold a government-accredited qualification.

## Marine safety and security

A key element in the maintenance of marine security is the availability of marine inspectors. The Directorate General of Marine Transport currently employs 242 type A marine inspectors (authorized to inspect ships of 500 gross tonnes and over), 360 type B marine inspectors (authorized to inspect ships of under 500 gross tonnes) and 53 radio inspectors, spread across the whole of Indonesia. They are required to attend regular refresher courses and meetings so that they can remain up to date with new developments in the field.

A second key element is the harbourmaster, the government official with the highest level of authority to monitor and enforce compliance with regulations on marine safety and security and the protection of the marine environment in a specific port. Indonesia has 287 harbourmasters across the archipelago.

The duties of a harbourmaster are to monitor ship seaworthiness; safety, security and order in the port; shipping traffic and the marine channel; loading and unloading activity; underwater activity and salvage; ship towing activity; piloting; loading and unloading of dangerous goods and dangerous or poisonous waste; fuel filling and fuel bunkers; embarkation and disembarkation; dredging and reclamation; and port facility development. The harbourmaster is also responsible for executing search and rescue assistance; minimizing lead pollution and extinguishing fires in the port; and monitoring the implementation of marine environment protection measures.

To perform these tasks and functions, the harbourmaster has the authority to coordinate all governance activity in the port; inspect and file letters, documents and ship news; publish ship activity agreements; conduct ship inspections; publish marine agreements; conduct ship accident inspections; detain ships on behalf of the ruling court; and conduct seafarer reviews and approvals.

A third element in the reform of marine safety administration has been the establishment of a special body to investigate shipping accidents, the Maritime Court (Mahkamar Pelayaran). Under the International Convention for the Safety of Life at Sea (SOLAS),[7] adopted in 1974, Indonesia is obliged to investigate accidents that take place in its seas, but until the establishment of this court it had no formal, standing means of conducting such investigations. The legal mandate of the Maritime Court is to identify the causes of shipping accidents and to apply administrative sanctions on ship masters and officers if they are found to be at fault. Its investigations are to be presided over by a council of at least five judges with the appropriate nautical and legal qualifications.

The sanctions that can be applied by the Maritime Court on those found guilty of causing an accident through error or carelessness range from issuing a warning to revocation of a marine certificate for a maximum of two years. Structurally the Maritime Court is an executive body of the Ministry of Transport, not a general judicial body, despite its judicial duties. This is because its main purpose is to promote good conduct.

---

7   The text of the convention is available at http://www.imo.org/Conventions/contents.asp?topic_id=257&doc_id=647.

Decisions of the Maritime Court are final and proceedings are open to the public. The legal foundation of the Maritime Court is provided by Law No. 17/2008, Presidential Decree No. 1/1998 and Ministerial Decree No. 55/2006. Several chapters of the Shipping Law expand the authority of the Maritime Court to investigate shipping accidents that involve collisions between commercial vessels; collisions between commercial vessels and state vessels; and collisions between commercial vessels and war vessels. In general, the court investigates all shipping accidents that occur in Indonesian waters or that involve Indonesia-flagged vessels outside Indonesia. It has no authority to punish foreign ship masters or foreign-flagged vessels. In such cases, the findings of an investigation will be passed to the country that has issued the certificate for follow-up.

Before a case can be handed to the Maritime Court, the Directorate General of Marine Transport must conduct a preliminary investigation and decide whether to proceed with the case. Thus, not all ship accidents are investigated by the Maritime Court, which handles about 50 cases a year. The goal of the Indonesian government is to make sure that every shipping accident is investigated and evaluated. Based on Presidential Decree No. 3/2007, a national safety evaluation and transport security team has been established to conduct research on, and suggest improvements to, regulations, standards and existing transport safety procedures. In addition, based on Presidential Decree No. 105/1999, the National Transport Safety Committee has operated since 1999 as a non-structural body inside the Ministry of Transport. Its main function is to investigate and study all transport accidents in order to prevent a recurrence of similar incidents.

The high number of marine accidents in recent years suggests that much remains to be done to improve the standard of marine safety in Indonesia. Nonetheless, Indonesia's determination to make progress in this area has been recognized with its re-election to the executive board of the IMO Council in 2007 as a category (c) board member, that is, one with a special interest in 'maritime transport or navigation, and whose election to the Council will ensure the representation of all major geographic areas of the world'. Indonesia has also embarked on a program of cooperation with the IMO to establish a consultative forum that will bring together government bodies and a broad range of stakeholders to consider issues related to safety at sea. The aim is to identify measures to reduce the incidence of ferry accidents in Indonesian waters. Indonesia has also been cooperating with Australia through the Indonesia Transport Safety Assistance Package (ITSAP), which was initiated in early 2008. One of the most important areas of cooperation is a project to establish standards for non-conventional vessels.

## CONCLUSION

Maintaining a high standard of marine safely is critically important for an archipelagic state such as Indonesia. Not only does the welfare of travellers and seafarers need to be safeguarded, but Indonesia's reputation as a safe destination needs to be preserved. Marine security comprises many elements, including navigation, ports, transport, monitoring, security and environmental protection. Handling these tasks is the responsibility of the Indonesian government through the Directorate General of Marine Transport. Indonesia has made considerable progress, but continues to face constraints due to a lack of funding, which restricts the availability of equipment, human resources and training.

## REFERENCES

Boisson, Philippe (1999), 'The history of safety at sea', available at http://www.oceansatlas.com/unatlas/issues/safety/transport_telecomm/history_safety/history_safety.htm, accessed 3 December 2008.

Fidrus, Multa (2008), 'Governor tells Merak Port to stop operating old vessels', Jakarta Post, 14 October, available at http://www.thejakartapost.com/news/2008/10/14/governor-tells-merak-port-stop-operating-old-vessels.html, accessed 1 December 2008.

Ho, Joshua H. (2006), 'The security of sea lanes in Southeast Asia', Asian Survey, 46(4): 558–74.

Nurhayati, Desy (2007), 'Coast guard argument continues', Jakarta Post, 22 October, available at http://www.thejakartapost.com/news/2007/10/22/coast-guard-argument-continues.html, accessed 3 December 2008.

Rutz, Werner O.A. and James R. Coull (1996), 'Inter-island passenger shipping in Indonesia; development of the system: present characteristics and future requirements', Journal of Transport Geography, 4(4): 275–86.

Sijabat, Ridwan Max (2007a), 'Toyo Fuji vows to comply with safety regulations' Jakarta Post, 10 April, available at http://www.thejakartapost.com/news/2007/04/10/toyo-fuji-vows-comply-safety-regulations.html, accessed 3 December 2008.

Sijabat, Ridwan Max (2007b), 'RI told to crack down on FOC cargo vessels', Jakarta Post, 19 May, available at http://www.thejakartapost.com/news/2007/05/19/ri-told-crack-down-foc-cargo-vessels.html-0, accessed 3 December 2008.

Soedarjo, Alvin Darlanika (2007), 'Graft often compromises marine safety, experts say', Jakarta Post, 6 January, available at http://www.thejakartapost.com/news/2007/01/06/graft-often-compromises-marine-safety-experts-say.html, accessed 3 December 2008.

Taufiqurrahman, M. (2007), 'Transport safety body wants more independence', Jakarta Post, 16 January, available at http://www.thejakartapost.com/news/2007/01/16/transport-safety-body-wants-more-independence.html, accessed 1 December 2008.

# 10 GOVERNANCE IN INDONESIA'S MARINE PROTECTED AREAS: A CASE STUDY OF KOMODO NATIONAL PARK

*Rili Djohani*

---

The governance of marine protected areas (MPAs) in Indonesia has undergone major changes over the past half-century. The system of centralized, technocratic management used in the 1970s and 1980s has since given way to a more community-focused approach. In the early 1990s, a series of natural resource management programs inspired by the community-based protected area programs in the Philippines were initiated with international support (White, Alino and Meneses 2006). Several collaborative management programs were subsequently established in support of the national parks and large-scale protected areas that form the basis for networks of MPAs across the Indonesian archipelago and in the Coral Triangle. These programs explicitly sought to address the problem of limited participation of local people that characterized the earlier system (TNC et al. 2008).

Using Komodo National Park as a case study, this chapter describes the collaborative management regime in Indonesian MPAs. It examines the ways in which government institutions and groups of resource users have shared responsibility for the park's management, as well as the influence of external factors on the governance and performance of the park. The chapter sheds light on the underlying assumptions and challenges associated with the implementation of collaborative management practices in Komodo National Park.

## INDONESIA'S MARINE PROTECTED AREAS

Healthy marine resources require healthy, intact ecosystems. Productive marine and coastal ecosystems are a source of goods and services that support communities and economies, including food security, tourism opportunities and coastal protection. They also help to maintain the full range of genetic variation that is essential to securing viable populations of key species, to sustaining evolutionary processes and to ensuring resilience in the face of natural and human disturbances (Agardy and Staub 2006; IUCN World Commission on Protected Areas 2008).

The Dutch colonial government established the first MPAs in the Indonesian archipelago, usually small areas of 1–2 hectares such as the Pulau Pombo reserve in Maluku. In the 1970s and 1980s, the central government undertook a massive expansion of national parks and nature reserves under the auspices of the Ministry of Forestry's Directorate General of Forest Protection and Nature Conservation. Most of the national parks created during this period were on land, but some, such as Komodo National Park, covered marine areas. In response to the growing recognition of the need for conservation of the marine environment, the government has since almost doubled the number of MPAs. In total, Indonesia now has 95 MPAs covering over 13 million hectares (see Table 10.1).[1] This already exceeds the government target of 10 million hectares by 2010 and puts it on track to protect 20 million hectares by 2020. At the May 2009 World Ocean Conference, the Indonesian government announced Southeast Asia's largest MPA: the Savu Sea Marine National Park covering 3,500,000 hectares, established by the Ministry of Marine Affairs and Fisheries.

An important characteristic of the second generation of MPAs has been the establishment of networks connecting small and large MPAs in recognition of the fact that, in isolation, small areas cannot support sustainable populations of fish and invertebrates. In many regions, economic, social and political constraints have made it impractical to create single MPAs of sufficient size to support viable, self-sustaining, diverse populations of marine species. There is international recognition in such

---

1 Two ministries have a mandate to establish protected areas in Indonesia: the Ministry of Forestry and the Ministry of Marine Affairs and Fisheries. Each is responsible for four categories of MPA. In the case of the Ministry of Forestry, the Directorate General of Forest Protection and Nature Conservation (Ditjen Perlindungan Hutan dan Konservasi Alam, or PHKA) exercises jurisdiction (under Law No. 5/1990). In the case of the Ministry of Marine Affairs and Fisheries, responsibility is exercised by the Directorate General of Marine, Coastal and Small Island Affairs (Ditjen Kelautan, Pesisir dan Pulau Pulau Kecil, or KP3K) (under Government Regulation (PP) No. 60/2007).

Table 10.1  Marine protected areas in Indonesia

| Category | Number of MPAs | Total area (hectares) |
|---|---|---|
| **Directorate General of Forest Protection and Nature Conservation, Ministry of Forestry (PHKA)** | | |
| Marine wildlife reserve (*suaka marga satwa*) | 7 | 339,218 |
| Marine nature reserve (*cagar alam laut*) | 9 | 274,215 |
| Marine national park (*taman nasional laut*) | 7 | 4,045,049 |
| Marine recreation park (*taman wisata laut*) | 18 | 767,610 |
| **Directorate General of Marine, Coastal and Small Island Affairs, Ministry of Marine Affairs and Fisheries (KP3K)** | | |
| District MPA (*kawasan konservasi laut daerah*) | 24 | 3,155,572 |
| Community-based MPA (*kawasan konservasi perairan berbasis masyarakat*) | 26 | 2,085 |
| Fisheries reserve (*suaka perikanan*) | 3 | 453 |
| Marine national park (*taman nasional perairan*) | 1 | 3,500,000 |
| **Total** | **94** | **8,584,202** |

Source: Ministry of Forestry; Ministry of Marine Affairs and Fisheries.

circumstances that networks of small and moderately sized MPAs may help to reduce the harmful effects of human activity without compromising conservation and fishery benefits (PISCO 2007). If an MPA, or network of MPAs, is resilient, it should be able to withstand environmental fluctuations, recover from catastrophes and support populations that may be able to replenish other damaged populations (West and Salm 2003).

The marine national parks, national parks with marine extensions and marine recreation parks established in the late 1970s and early 1980s, such as Thousand Islands (Java), Komodo (East Nusa Tenggara), Ujung Kulon (southwest Java) and West Bali, were much larger in size than their antecedents. In the present century, marine ecoregions that cross national borders have also been established, such as the Sulu–Sulawesi network of marine ecoregions to the north of Kalimantan and Sulawesi, which are managed by Indonesia, the Philippines and Malaysia. Other areas have been protected with an eye to climate change, such as the ecoregions in the Lesser Sundas (comprising Bali, Lombok, Sumbawa, Sumba, Flores, Alor Solor, and East and West Timor) and Raja Ampat (comprising four main and many smaller islands off West Papua).

Most recently, the six Coral Triangle countries — Indonesia, the Philippines, Malaysia, Timor-Leste, Papua New Guinea and the Solomon Islands — have participated in a joint initiative to save their shared coral reefs, recognized as the world's centre of marine biodiversity. These interconnected coral reef ecosystems are a refuge for fish and coral as well as sustaining the livelihoods of millions of people. Under the Coral Triangle Initiative, the six governments have developed a series of regional and national action plans to protect the reefs, with support from international non-government organizations (NGOs) and donors. A declaration to this effect was signed by the six heads of state at a Coral Triangle summit held in Manado in May 2009.

## COLLABORATIVE MANAGEMENT OF MARINE PROTECTED AREAS

The Dutch established very simple governance structures based on strict controls under which no activity was allowed in a marine reserve. The much larger marine nature reserves (*cagar alam laut*) established in the 1980s and 1990s, such as Teluk Bintuni in West Papua, were also subject to strict control by the central government. But as the size and complexity of MPAs and MPA networks has increased, so too have their governance structures evolved from tight control by a single actor to more complex layers of control by a greater number of actors managing multiple-use areas in large MPAs.

In the 1990s, the government sought to consolidate governance in MPAs by opening up management processes to stakeholder involvement. This led to the emergence of several community-based initiatives across the country. Proyek Pesisir (the Coastal Resources Management Project), for example, was implemented between 1996 and 2003 in North Sulawesi, East Kalimantan and Papua to decentralize and strengthen natural resource management in Indonesia. Jointly funded by the United States Agency for International Development (USAID) and the National Development Planning Agency (Badan Perencanaan Pembangunan Nasional, or Bappenas), its goal was to guide development to achieve a better balance between the quality of life of coastal communities and the condition of Indonesia's coastal resources. The Coral Reef Rehabilitation and Management Program (Coremap) was another long-term program initiated by the government of Indonesia. It sought to protect, rehabilitate and promote the sustainable use of coral reefs and their associated ecosystems, and in doing so to enhance the welfare of coastal communities. These programs and others like them were developed in response to the lack of effective management and avenues for participation in the parks run by the central government. These parks were sometimes called 'paper parks' because, in effect, they existed only on paper.

After the Asian financial crisis hit Indonesia in 1997 and the New Order collapsed, the government sought international assistance for a number of collaborative management initiatives in national parks. These projects included financial support by USAID for a co-management initiative for Bunaken Marine National Park (off North Sulawesi) in 1998;[2] the Worldwide Fund for Nature's funding of a collaborative management regime for the Kayan Mentarang National Park (in the interior of Kalimantan) in 1999; and the support of The Nature Conservancy (TNC) for the establishment of a self-financing scheme and collaborative management initiative for Komodo National Park, also in 1999. The Directorate General of Forest Protection and Nature Conservation, the Worldwide Fund for Nature, Conservation International and TNC also joined forces to encourage stakeholders throughout Indonesia to have their say on a proposed new government regulation on collaborative management in protected areas. The Ministry of Forestry's new policy on collaborative management actively supported coalitions of interested groups in protected areas. In 2005, an Indonesian delegation presented this new policy, supported by case studies of collaborative management in Indonesia, at the seventh Conference of the Parties to the Convention on Biodiversity in Kuala Lumpur and the World Conservation Forum in Bangkok. More large MPAs and MPA networks have since been established in Berau (East Kalimantan), Wakatobi (Southeast Sulawesi), Raja Ampat (West Papua) and the Lesser Sundas (East Nusa Tenggara) with central government, local government, community, and local and international NGO involvement.

These developments reflect a global trend to recognize the importance of partnerships to support national parks and protected areas and, consequently, the influence of institutional and governance arrangements on the design, implementation and longevity of MPA networks. Governance refers to the institutional arrangement of actors across several organizational layers (Wilson 1989). Once it is acknowledged that governance occurs at different levels, it becomes possible to develop a site-specific framework within which to analyse the resilience of MPA

---

2  The objective of the Bunaken Natural Resources Management Project was to promote cooperation between the central government, the private sector, NGOs, civil society and other stakeholders to achieve three interrelated goals: (1) to clarify the roles and responsibilities of various groups with regard to natural resource decision making and management in the new, decentralized political environment; (2) to improve the capacity of local stakeholders to manage natural resources; and (3) to broaden public understanding of the necessity for sustainable natural resource management.

governance and performance in relation to the influence of various internal and external actors and factors.

The resilience perspective is increasingly used as a tool to increase understanding of the dynamics of social–ecological systems. Resilient systems are able to cope with, adapt to, shape and learn to live with uncertainty and surprise (Brand and Jax 2007). Non-resilient systems, in contrast, are prone to irreversible change and are at risk of shifting into another, often undesirable, state (Marshall and Marshall 2007). When applied to the environment, this approach focuses on the ability of an ecosystem to absorb or recover from disturbance and change while maintaining its functions and services. Coral reefs, for example, have proven to be more resilient to disturbance and stress when the reefs exhibit high coral recruitment, diversity and coverage, and support corals characterized by a broad range of sizes and ages (Salm, Done and McLeod 2006).

Governance resilience, too, can be defined as the ability of a governance structure to absorb or recover from disturbance and change while maintaining its functions and services. An MPA's governance structure can be considered resilient when it is able to absorb or adapt to major external and internal changes and disturbances without losing its ability to manage natural resources effectively. Governance here refers to the structures and processes through which social actors, including government, local communities, the private sector and NGOs, share power, develop decision-making processes and erect institutional arrangements. Laws, regulations, debates, negotiation, mediation, conflict resolution and public consultation are all part of the governance process. Emerging from these interactions, governance takes form through laws, regulations and other formal decision-making mechanisms, or is expressed informally by influencing the agendas and shaping the contexts in which decisions are contested and access to resources determined (Lebel et al. 2006). As Hyden, Court and Mease (2004) point out, it is important to make a distinction between governance *performance* indicators and governance *process* indicators. In the context of MPAs, performance indicators refer to the level of protection of natural resources or the extent of benefit for local people, whereas process indicators refer to the quality of governance in terms of how outcomes are achieved. Although there has been considerable improvement in monitoring MPA governance outcomes, the monitoring of processes has been limited. The challenge, therefore, is to measure governance cohesively and systematically in terms of critical processes. This can be very difficult considering the many different forms MPA governance may take.

In general, there are four main types of protected area governance: government management; private protection; community-based management; and collaborative management (Borrini-Feyerabend 2007). In their typology of MPA management approaches, Christie and White (2007) argue that centralized management is historically the most common governance regime in colonial and post-colonial societies. In the 1990s, community-based management strategies were frequently employed in remote areas with strong local tenure. Private management of Mpas is rare. Collaborative management can be considered a hybrid of bottom-up and centralized management in that it involves various formal and informal actors at multiple levels in MPA management.

As an institutional form, co-management involves the sharing of rights and responsibilities for a particular resource, thus encouraging a multi-level perspective (Berkes, George and Preston 1991). It can accommodate various levels of institutional partnerships at the state and non-state levels, and across local, national and even international boundaries. By providing mechanisms for the sharing of vision, resources, expertise and systems, at its best co-management is able to enhance the impact of natural resource management at various levels and on various scales. The co-management literature provides many examples of successful multi-level governance arrangements (Pinkerton 1989; Nadasdy 2003; Wilson, Nielsen and Degnbol 2003).

The levels of partnership between government, NGOs and private organizations can be described as consultative, coordinative, complementary, collaborative and critical (Pomeroy 1994, 1995; Fernandez 2007). In co-managed protected areas, authority, responsibility and accountability are shared in a variety of ways among a range of actors, likely to include one or more government agencies, local communities, private landowners and other stakeholders. Formal decision-making authority, responsibility and accountability still rests with one agency (often a national government agency), but the agency can by law or policy collaborate with other actors (Borrini-Feyerabend et al. 2004). To be effective, all the actors involved must recognize the legitimacy of their respective entitlements to manage the protected area. Appropriate legislative and regulatory frameworks are fundamental to the effective management of MPAs and MPA networks, as a poorly integrated array of legal and institutional responsibilities can lead to problems such as competing mandates, overlaps, gaps and inefficiencies, all of which hinder effectiveness. Although many countries have specific legislation establishing individual MPAs, and a variety of agencies with marine responsibilities, few have strategic legislative frameworks or institutional arrangements to support collaborative management in protected areas.

A number of factors have increased the influence of the collaborative management approach in Indonesia. In particular, the establishment of the Ministry of Marine Affairs and Fisheries, including a conservation section, in 1999 led to the rapid establishment of new types of MPAs to be managed collaboratively by local governments and communities. In the ensuing decade, the ministry established more MPAs than the Directorate General of Forest Protection and Nature Conservation had in the preceding four decades. Financial and political factors also played a part in the move to adopt collaborative management models: the Asian financial crisis of 1997–98 increased the dependence of the central government on foreign aid; and the decentralization policy adopted in 1999 further shifted authority to manage protected areas to district governments and other partners. From the late 1990s, aid agencies began to provide support directly to the district governments, forcing the forestry and fisheries ministries to consider more collaborative management approaches in marine national parks and district MPAs. It was in this context that the Ministry of Forestry joined forces with USAID, the Worldwide Fund for Nature and TNC to set up collaborative management programs with improved levels of governance in Bunaken, Kayan Mentarang and Komodo, as mentioned earlier. The Ministry of Marine Affairs and Fisheries, meanwhile, worked with the Worldwide Fund for Nature, TNC and Conservation International to support the establishment of district MPAs such as Derawan (off the eastern coast of Kalimantan) and Raja Ampat, drawing on the best scientific information on MPA design available at the time.

It is too early to assess whether the attributes of shared governance and collaborative management in these MPAs have led to an enhanced capacity to manage resilience. This question will be examined further in relation to the collaborative management initiative in Komodo National Park.

## THE KOMODO COLLABORATIVE MANAGEMENT INITIATIVE

Komodo National Park was established in 1980. In 1991 it was added to the World Heritage List, where it was described as 'One of the most important and significant natural habitats for in-situ conservation of biological diversity, including those containing threatened species of outstanding universal value from the point of view of science or conservation'. In 1997 it was declared a Man and Biosphere Reserve. The Komodo National Park is universally recognized as an outstanding store of globally significant terrestrial and marine biodiversity. It is situated in the heart of the Coral Triangle, an area known to scientists and conservationists as having the richest coral reef biodiversity in the world. It is also located in the Wallacea bioregion, an area of great importance for terrestrial conservation.

Since 1995, the Komodo National Park has received support from TNC for surveillance, outreach, monitoring and alternative livelihood programs. The park was one of the few protected areas in Southeast Asia to actually record an increase in coral coverage in the period 1996–2002, mainly through a significant reduction in dynamite fishing (Mous et al. 2003). During this period, the Directorate General of Forest Protection and Nature Conservation, with the support of TNC, designed a 25-year management plan for the park based on extensive stakeholder consultations, socio-economic studies and ecological surveys. Adopted by the central and local governments in 2000, this plan now constitutes the legal framework for the regulation of resource use in the park's protection and wilderness zones, exclusive user zones, tourism zones and pelagic fishery zones. The overall goal of the plan is to 'protect [the park's] biodiversity, both the Komodo dragon and the breeding stocks of commercial fishes for replenishment of surrounding fishing grounds', by reducing both the threats to the area's resources and the conflicts between incompatible activities (PKA and TNC 2000a, 2000b, 2000c).

Financing the implementation of the 25-year management plan proved difficult in the wake of the financial crisis, which caused cutbacks in government conservation budgets of over 80 per cent. In response, the Directorate General of Forest Protection and Nature Conservation and TNC developed a scheme to use the revenues generated by tourism to fund the management and maintenance of the park, while also returning benefits to the people who depended on its natural resources for their livelihoods. They proposed a collaborative management structure in which the local government and local stakeholders — communities, NGOs, entrepreneurs and so on — would be given representation on a consultative community forum.

With the support of the Global Environmental Facility and the International Finance Corporation, in 2002 TNC established PT Putri Naga Komodo (PNK), a joint venture company that would invest in infrastructure, marketing, capacity building, conservation and community development programs with the aim of making the park financially self-sufficient by 2012. The tourism revenues collected by PNK were to be reinvested to support conservation, the local people and continuing efforts to improve the visitor experience in the park, including better-trained guides, more hiking tracks, educational interpretation panels and hospitality services. Between its introduction in May 2006 and December 2007, the conservation fee levied on visitors raised US$427,000, with 2007 revenues jumping 23 per cent over the previous year. After virtually disappearing in the late 1990s, the cruise market bounced back, with 10 ships calling at Komodo in 2007 and cruise ship passengers accounting for a third of all arrivals.

The Komodo Collaborative Management Initiative signed in 2003 between the Komodo National Park Authority and PNK set out the roles and responsibilities of each party. Under the agreement, the park authority remained responsible for park management and enforcement, while PNK became responsible for generating revenue, managing marketing and improving tourism facilities and products in the park. In addition, the company was to fund conservation and community development programs. In 2004, PNK obtained a 30-year tourism licence that allowed it to develop and manage visitor facilities in the concession area, collect tourism service charges and establish a collaborative management structure for the park. The park in turn has benefited from the training, resources and technical expertise on conservation, tourism and community development provided by the company.

With the establishment of the new district of West Manggarai on the nearby island of Flores in 2004, the park became the focus of competing political agendas. Issues of access and exploitation have frequently been raised in the local parliament—although the underlying issues have more to do with local feuds between various groups in the district. On the other hand, the district heads elected since 1995 have consistently supported the Komodo Collaborative Management Initiative, motivated by the need to preserve the natural ecosystems that are critical for the local economy, which is heavily dependent on fisheries and tourism. Although decentralization has politicized park affairs, the consistently strong alliance between PNK and the office of the district head has ensured continuing local support.

A number of conservation strategies and management interventions have been initiated by the park authority and PNK in partnership with local stakeholders and communities. These include research and monitoring to measure the health of corals and reef fish communities; measures to assess the resource-use patterns of residents living in and around the park; and ways to improve surveillance and enforcement of the park's zoning system, which regulates access to and use of the park, both on land and at sea. A small fleet of 'floating ranger stations' is now providing a more permanent presence in the park and has considerably reduced destructive fishing practices—although the overfishing of vulnerable sites by traditional fishing methods still constitutes one of the conservation program's greatest challenges. Also important have been educational programs aimed at raising the awareness of local communities and other stakeholders about the importance of conservation, and targeted alternative income projects developed in conjunction with microfinancing schemes. In one microfinance-based cooperative, local women are making Komodo dragon bracelets from coconut beads for sale to tourists. Like two other recipients of microcredit, a small bakery and a workshop where fishing gear and fishing boats are repaired, this cooperative has already repaid close to 100 per cent of its loan. In 2007, the district head of West Manggarai established a community-based advisory body, the Forum for Community Communications, to help resolve conflicts involving the park. This was done in anticipation of a formal board being set up for the Komodo Collaborative Management Initiative, where local government, community and stakeholder representatives would be able to provide feedback to the park authority.

The park's collaborative management structure has responded well to external factors such as the global financial crisis by mobilizing international support. The park's tourism-based self-financing structure has also proved resilient in difficult economic conditions. These successes notwithstanding, there have been a number of internal and external problems related to the governance of the park. A high turnover of staff in the Directorate General of Forest Protection and Nature Conservation, and within the park authority, has made it difficult to coordinate and implement the rules developed collaboratively by the park, PNK and the local government over time. The joint venture also faces challenges in retaining senior on-site managers. Moreover, neither the park authority nor the local government is willing to go ahead with the formal establishment of a board for the Komodo Collaborative Management Initiative until the conflicts between government and local communities over the use of park resources are resolved. These include the right to collect bird's nests in the park, rights over the ownership of wild horses and land, and the right of local people to exploit park land for agriculture.

Fundamental disagreements between the partners have brought previously agreed rules into question. In particular, the high turnover of senior staff in the Directorate General of Forest Protection and Nature Conservation has made it very difficult to ensure continuity of management, which depended heavily on individual leadership at different levels within the agency. In some cases the collaborative management group has responded well to external actors (such as fishermen using illegal and destructive fishing practices) and to criticism from national NGOs, community members and journalists. In other cases, however, it has failed to do so, because a lack of alignment between the park's executives, managers and operators has made it difficult for them to coordinate a coherent response.

## CONCLUSIONS

The experience of the Komodo Collaborative Management Initiative strongly suggests that coordination and cohesiveness among partners is crucial for the overall performance of an MPA and its ability to adapt to external shocks. During periods when the park authority, the local government and PNK agreed on basic principles of collaboration and approval of joint work plans, this synergy translated into strong implementation of programs and a strong constituency on the ground. Over time, though, it became clear that the agreed long-term framework for collaborative management, roles and responsibilities, and work plans could not be taken for granted, but would require constant communication, capacity building and adaptive management by actors at all levels. Particularly in the context of high turnover of staff, it has become clear that continuity is pivotal, not only in legal frameworks and agreements but also within the institutions and communities that are part of the collaborative management process in Komodo.

The experience so far suggests that the legal framework for the Komodo Collaborative Management Initiative should include a mechanism to address stakeholder grievances, in order to ensure the resilience of the MPA's governance structure. The Forum for Community Communications is an important step forward in this regard, but it needs to be strengthened if it is to become an effective platform for stakeholder discussions and communication. The benchmark for successful collaborative management in Komodo would be the formal establishment of a board for the Komodo Collaborative Management Initiative, on which representatives from the central and local governments, the park authority, the Forum for Community Communications and PNK would sit. Together, they could help set directions for the park and ensure the enforcement of its zoning system. The biggest challenges facing the park remain the enforcement of no-take zones and overfishing of reefs by fishermen using traditional fishing methods.

The collaborative management approach adopted by Komodo National Park has helped to protect the park from destructive fishing practices and build the capacity to monitor coral reef ecosystems. The park now has good infrastructure and well-trained local guides. A number of long-term community development programs have been implemented and a tourism fee system has been set up as part of a long-term financing plan for the park. The local government has issued a number of supportive local government regulations to help regulate fisheries.

The public–private partnership in Komodo has been successful in mobilizing resources to support conservation, community development and tourism, and in establishing a number of frameworks and systems that will support collaborative management and long-term financing. New national policies continue to be developed to support additional government fee systems in Indonesia's national parks. To strengthen the resilience of the park's management structure, the partners in Komodo must strengthen the relationship between the various stakeholders and agree on and institutionalize the basic principles for collaboration. If they do not, then the park's governance structure will remain a delicate balance of efforts that may not be resilient to major external and internal changes and disturbances in the long term.

## ACKNOWLEDGMENTS

The author is grateful to all those who have collaborated in the design and implementation of the Komodo Collaborative Management Initiative, from Ministry of Forestry and local government officials to members of the West Manggarai local community, The Nature Conservancy, PT Putri Naga Komodo, the International Finance Corporation, the private sector and local NGOs. I especially appreciate the input and comments of Dr Adriaan Bedner from the Van Vollenhoven Institute, Faculty of Law, University of Leiden; the encouragement of Dr Robert Cribb from the Australian National University; the suggestions of the book's copy editor, Beth Thomson; the invaluable support of Russell Leiman and Ian Duttton; and the comments of my colleagues in The Nature Conservancy and PT Putri Naga Komodo, in particular Abdul Halim, Johannes Subijanto, Widodo Ramono, Eleanor Carter, Alan White, Rod Salm, Joanne Wilson, Tri Soekirman and Marcus Matthews-Sawyer.

## REFERENCES

Agardy, T. and F. Staub (2006), 'Marine protected areas and MPA networks', Network of Conservation Educators and Practitioners, Center for Biodiversity and Conservation, American Museum of Natural History, New York, available at http://ncep.amnh.org/.

Berkes, F., P. George and R. Preston (1991), 'Co-management: the evolution of the theory and practice of joint administration of living resources', *Alternatives*, 18(2): 12–18.

Borrini-Feyerabend, G. (2007), 'The IUCN protected area matrix: a tool towards effective protected area systems', paper presented the Summit *on the IUCN categories*, IUCN World Commission on Protected Areas Task Force: IUCN Protected Areas Categories, Andalusia, 7–11 May.

Borrini-Feyerabend, G., M. Pimbert, M.T. Farvar, A. Kothari and Y. Renard (2004), *Sharing Power: Learning-by-doing in Co-management of Natural Resources throughout the World*, IIED and IUCN/CEESP, Cenesta, Tehran.

Brand, F.S. and K. Jax (2007), 'Focusing the meaning(s) of resilience: resilience as a descriptive concept and a boundary object', *Ecology and Society*, 12(1): 23.

Christie, P. and A.T. White (2007), 'Best practices for improved governance of coral reef marine protected areas', *Coral Reefs*, 26: 1,047–56.

Fernandez, P.R. (2007), 'Understanding relational politics in MPA governance in northeastern Iloilo, Philippines', *Journal of Coastal Research*, special issue, 50: 38–42.

Hyden, Goran, Julius Court and Kenneth Mease (2004), *Making Sense of Governance: Empirical Evidence from 16 Developing Countries*, Lynne Rienner Publishers, Boulder CO.

IUCN World Commission on Protected Areas (IUCN–WCPA) (2008), *Establishing Resilient Marine Protected Area Networks: Making it Happen*, IUCN–WCPA, National Oceanic and Atmospheric Administration and The Nature Conservancy, Washington DC.

Lebel, Louis, John M. Anderies, Bruce Campbell, Carl Folke, Steve Hatfield-Dodds, Terry P. Hughes and James Wilson (2006), 'Governance and the capacity to manage resilience in regional social–ecological systems', *Ecology and Society*, 11(1): 19.

Marshall, N.A. and P.A. Marshall (2007), 'Conceptualizing and operationalizing social resilience within commercial fisheries in northern Australia', *Ecology and Society*, 12(1): 14.

Mous, P.J., J.S. Pet, D. Gede Raka, J. Subijanto, A.H. Muljadi and R.H. Djohani (2003), 'How monitoring demonstrated effective control of blast fishing in Komodo National Park', in C. Wilkinson, A. Green, J. Almany and S. Dionne (eds), *Monitoring Coral Reef Marine Protected Areas*, Global Coral Reef Monitoring Network and Australian Institute of Marine Science, Townsville, pp. 20–21.

Nadasdy, P. (2003), *Hunters and Bureaucrats*, UBC Press, Vancouver.

Pinkerton, E. (ed.) (1989), *Cooperative Management of Local Fisheries, New Directions for Improved Management and Community Development*, University of British Columbia Press, Vancouver.

PISCO (Partnership for Interdisciplinary Studies of Coastal Oceans) (2007), *Science of Marine Reserves*, second edition, international version, available at http://www.piscoweb.org.

PKA and TNC (Direktorat Jenderal Perlindungan dan Konservasi Alam and The Nature Conservancy) (2000a), *25 Year Master Plan for Management Komodo National Park, Book 1: Management Plan*, Jakarta, available at http://www.komodonationalpark.org/.

PKA and TNC (Direktorat Jenderal Perlindungan dan Konservasi Alam and The Nature Conservancy) (2000b), *25 Year Master Plan for Management Komodo National Park, Book 2: Data and Analysis*, Jakarta, available at http://www.komodonationalpark.org/.

PKA and TNC (Direktorat Jenderal Perlindungan dan Konservasi Alam and The Nature Conservancy) (2000c), *25 Year Master Plan for Management Komodo National Park, Book 3: Site Planning*, Jakarta, available at http://www.komodo nationalpark.org/.

Pomeroy, R.S. (ed.) (1994), 'Community management and common property of coastal fisheries in Asia and the Pacific: concepts, methods and experiences', in *ICLARM Conference Proceedings 45*, International Center for Living Aquatic Resources Management (ICLARM), Manila, pp. 124–44.

Pomeroy, R.S. (1995), 'Community-based and co-management institutions for sustainable coastal fisheries management in Southeast Asia', *Ocean and Coastal Management*, 27(3): 143–62.

Salm, R.V., T. Done and E. McLeod (2006), 'Marine protected areas: planning in a changing climate', in J.T. Phinney, O. Hoegh-Guldberg, J. Kleypas, W. Skirving and A. Strong (eds), *Coral Reefs and Climate Change: Science and Management*,

Coastal and Estuarine Studies 61, American Geophysical Union, Washington DC, pp. 207–21.

TNC, WWF, CI and WCS (The Nature Conservancy, World Wildlife Fund, Conservation International and Wildlife Conservation Society (2008), *Marine Protected Area Networks in the Coral Triangle: Development and Lessons*, TNC, WWF, CI, WCS and the United States Agency for International Development, Cebu City.

West, J.M. and R.V. Salm (2003), 'Resistance and resilience to coral bleaching: implications for coral reef conservation and management', *Conservation Biology*, 17(4): 956–67.

White, A.T., P.M. Alino and A.T. Meneses (2006), *Creating and Managing Marine Protected Areas in the Philippines*, Fisheries Improved for Sustainable Harvest Project, Coastal Conservation and Education Foundation, Inc. and University of the Philippines Marine Science Institute, Cebu City.

Wilson, J.Q. (1989), *Bureaucracy: What Government Agencies Do and Why They Do It*, Basic Books, New York.

Wilson, D.C., J.R. Nielsen and P. Degnbol (eds) (2003), *The Fisheries Co-management Experience: Accomplishments, Challenges and Prospects*, Kluwer Academic Publishers, Dordrecht.

# 11 RISING TO THE CHALLENGE OF PROVIDING LEGAL PROTECTION FOR THE INDONESIAN COASTAL AND MARINE ENVIRONMENT

*Sarah Waddell*

Indonesia's marine resources have been severely affected by human actions such as overfishing, destructive fishing methods, clearing of mangrove forests and uncontrolled discharge of industrial and domestic waste into rivers and the sea (State Minister for the Environment 2006: Ch. 7). Threats to fish and other biota as well as coral reefs and seagrass beds come from both sea and land-based human activity. Indonesia does not have a strong record of adopting and implementing measures to address these threats. Lack of administrative and financial resources is often given as the main reason for weak implementation but obstacles have also arisen from the legal arrangements for environmental and natural resource management. Indeed, the various shortcomings of Law No. 23/1997 on Environmental Management have been discussed for many years by Indonesian environmental lawyers.

Two relatively new laws to govern the exploitation of coastal and marine resources offer a new and more promising approach. Law No. 31/2004 on Fisheries and Law No. 27/2007 on the Management of Coastal Areas and Small Islands offer the possibility of an ecologically based approach to managing fisheries and coastal areas backed by stronger sanctions. The laws mark a move away from maximizing levels of production and exploitation towards management for sustainability. They place a legal requirement on the Minister for Marine Affairs and Fisheries to prepare management plans for the sustainable exploitation of fishery resources and to determine the extent and potential of the country's fish resources. They combine new obligations on government with a more comprehensive range of sanctions for breaches of the provisions to protect fisheries and coastal resources. Moreover, as a result of the Law on Coastal and Small Island Management, areas that were previously not part of spatial planning have been brought into a planning and management regime that acknowledges the need to set functional limits on the exploitation of resources that take account of carrying capacity and ecological processes.

This chapter will start by reviewing some of the legal shortcomings that have been identified in environmental management to date. It will then highlight important new legal developments in coastal and marine resources management but also consider, from a legal perspective, weaknesses that could impede effective implementation.

## THE STATE OF THE MARINE AND COASTAL ENVIRONMENT

### The impact of fishing activity on fish stocks and habitats

In the boom-and-bust history of fisheries development in Indonesia, fish stocks have been declining steadily, with the first concerns about the status of stocks being raised in the 1930s. It is clear that over the past 40 years many fish populations have been severely depleted.[1] The official assessment is that Indonesia is still within its catch limit,[2] but various attempts to determine the magnitude of the remaining resources reveal major discrepancies (Gillet 1996). Information provided by the Ministry of Marine Affairs and Fisheries categorizes fish stocks in Indonesian seas as 'uncertain', 'overfished', 'fully exploited', 'moderate' or 'not available' (Table 11.1).

Estimating the exact extent of depletion is fraught with difficulty, as information on number and size of boats, catch size and fishing gear is incomplete. For example, the available statistics do not provide details of fish catch but only of fish landed in Indonesia. Fish that are exported without landing, by foreign, joint venture or national industrial-scale vessels with freezers, may bypass the recording system altogether. Moreover, landing figures do not give a reliable indication of the place of the catch, which may be a great distance from the home port of the fishing boats

---

1  See, for example, Butcher (2004: Chs 6 and 7), Butcher (2005: 1) and Morgan and Staples (2006).
2  The Food and Agriculture Organization (FAO) has set the catch limit at 80 per cent of the total potential yield of an area. The 2006 State Minister for the Environment report estimates that Indonesia's level of exploitation is 77.7 per cent of the total potential yield.

*Table 11.1    Status of fishery resource utilization in nine Indonesian fishery management zones*

| Location | Large pelagic | Small pelagic | Demersal | Shrimp |
|---|---|---|---|---|
| Malacca Strait | Uncertain | Fully exploited | Overfished | Overfished |
| South China Sea | Uncertain | Overfished | Fully exploited | Moderate |
| Java Sea | Uncertain | Overfished | Fully exploited | Fully exploited |
| Makassar Strait & Flores Sea | Uncertain | Moderate | Fully exploited | Overfished |
| Banda Sea | Moderate | Uncertain | Uncertain | Uncertain |
| Arafura Sea | Uncertain | Moderate | Fully exploited | Overfished |
| Tomini Gulf & Molucca Sea | Uncertain | Not available | Moderate | Moderate |
| Sulawesi Sea & Pacific Ocean | Overfished | Uncertain | Uncertain | Not available |
| Indian Ocean A (west of Sumatra) | Fully exploited | Moderate | Fully exploited | Fully exploited |
| Indian Ocean B (south of Java–Nusa Tenggara) | Fully exploited | Fully exploited | Fully exploited | Fully exploited |

*Source:* Presentation by Dr Aji Sularso, Director-General of Surveillance and Control on Marine Resources and Fisheries, Ministry of Marine Affairs and Fisheries, 2008, citing National Commission on Stock Assessment of Marine Fisheries Resources (2006).

(Fegan 1999: 5). It is therefore difficult to determine the fishing pressure on any specific ecological region in Indonesia with any real certainty.

One indicator of the likely pressure on resources is the growth of the Indonesian fishing fleet and changes in its character in recent times. In 1960, 170,000 fishing boats were officially recorded as operating in Indonesian waters; of these, only 1,500 had motors of any type (Krisnandhi 1969: 51). According to the most recent statistics from the Ministry of Marine Affairs and Fisheries, in 2005 the number of motorized boats had risen to 125,750 (cited in State Minister for the Environment 2006: Table 7.3). Artisanal or small commercial fishing operations are numerically dominant: small boats of 30 gross tonnes or less, including motorboats (*perahu motor tempel*), still make up 99 per cent of the fleet. However, the same source indicates that there are approximately 7,720 larger vessels fishing in Indonesian seas. They constitute less than 1 per cent of the fleet but are capable of catching far greater volumes of fish than the smaller vessels.

Threats to Indonesia's fish stocks come not only from the growing scale of fishing activity but also from the choice of fishing methods. In 1981, in recognition of the destructive effect of trawling, the government banned this form of fishing within most of Indonesia's exclusive economic zone (EEZ).[3] However, trawlers (*kapal pukat harimau*) were still allowed east of longitude 130°E, leading to a concentration of legal trawling on the wide, shallow shelf of the Arafura Sea running west from the southern coast of West Papua (Fegan 2003). Even within the area in which trawlers are banned, fishing boats using trawl-like gear, but called fish net boats (*kapal pukat ikan*), still operate both legally and illegally (Zamansyah 2000; Fegan 2003; Butcher 2005: 2–3).

Since the early 1980s, intensive offshore fishing by tuna longliners (mostly from Taiwan) and large-scale offshore industrial purse seining by Philippine boats have seriously reduced yellowfin tuna and other fish stocks in the waters off northern Sulawesi (Naamin, Mathews and Monintja 1995). Another destructive fishing method is the use of fine-mesh nets, which catch juvenile fish and other marine organisms that are not of any commercial value to the fishing industry but are essential for ecosystem sustainability. Blast fishing, in which explosives are used to stun marine life for easy collection by fishermen, and cyanide fishing, which achieves the same effect by poisoning an area of sea, are both highly destructive. Also considered to be destructive are portable traps (*bubu*) and the practice of *muroami* — the use of an encircling net together with pounding devices to catch fish (Pet-Soede and Erdmann 1998: 29).

Official statistics show increases in the use of various kinds of fishing gear from 2001 to 2005 (MMAF 2006: Table 1.5), some of which have been classified as destructive. Most notable were the increases in the use of trawling with stern shrimp trawls (*pukat tarik udang tunggal*) (279 per cent) and fish nets (*pukat tarik ikan*) (76 per cent). The use of tuna longlines had risen by 31.8 per cent, crab gear by 107 per cent, *bubu* by 100 per cent, cast nets by 76 per cent, lift nets (other forms) by 68 per cent and *muroami* by 61 per cent. Even if these figures cannot be regarded as precise, they point to a dramatic increase in fishing pressure on Indonesian fish stocks in a short space of time.

The true extent of capture fishing is also difficult to estimate due to the scarcity of data on illegal, unreported and unregulated (IUU) fishing. Statistics do not capture the activities of fishing vessels from China, Taiwan, Thailand, Vietnam, the Philippines and Malaysia operating illegally in Indonesian waters. As this fishing is illegal, it is also unregu-

---

3  Presidential Decree No. 39/1980 on the Eradication of Trawl Fishing.

lated, without any control over aspects such as fishing gear, catch size, minimum fish size or fishing practices (Sodik 2007; Palma and Tsamenyi 2008). The problem of IUU fishing is particularly serious in eastern Indonesia where surveillance is weak.[4]

### Threats from the land

Threats to fish stocks also come from the deteriorating condition of mangrove areas, which are the nurseries for larger marine fish species as well as being important direct sources of shrimp, crab and small fish. Mangroves occur naturally along many coastlines in the archipelago. According to data prepared by the Ministry of Forestry, Indonesia has 4,390,756 hectares of classified mangrove areas (*kawasan mangrove*). In these areas, 1,830,877 hectares (28 per cent) of mangrove forests have been destroyed; outside them, the rate of destruction is as high as 72 per cent (State Minister for the Environment 2007). Mangrove along the eastern side of Sumatra, which was once considered the most extensive in Indonesia, is now badly damaged. The loss of mangroves has been caused both by clearing for palm oil plantations and by unchecked land conversion for shrimp farming and fish farms.

In addition, urban areas like Jakarta and Surabaya produce massive amounts of untreated pollution from industrial and domestic sources, which is discharged directly into the sea. Cases of mass fish kills as a result of a sudden influx of pollution are reportedly on the increase.[5] Jakarta Bay is becoming progressively more eutrophic from very high nutrient concentrations. A recent biophysical assessment of the bay by the United Nations Educational, Scientific and Cultural Organization (UNESCO) identified three major sources of pollution: industrial, domestic and agricultural. The two types of industrial pollutants that have been identified in Jakarta Bay are heavy metals (copper, lead and mercury) and polychlorobiphenyls (PCBs). PCBs have been found in marine plankton, fish, mammals, birds and the human body. Ship waste, oil spillages and offshore mining are other sources of industrial pollution (UNESCO 2000; Arifin 2004).

---

4  To counter IUU fishing, in 2008 the Minister for Marine Affairs and Fisheries, Freddy Numberi, proposed the world's first Regional Plan of Action on responsible fisheries practices, to be based on the FAO's International Plan of Action (see Numberi 2008).

5  In 2005, for example, large quantities of fish in the Thousand Islands Marine Park were found dead, probably because of a drop in the oxygen content of the water ('Many fish die of asphyxiation', *Jakarta Post*, 9 August 2005).

## OBSTACLES ARISING FROM LEGAL PROVISIONS ON ENVIRONMENTAL MANAGEMENT

While the state of Indonesia's marine and coastal environment is a matter to be addressed by legislation governing fisheries management and the management of coastal and marine resources, some of the issues mentioned above fall within the purview of environmental law. Environmental law can, for example, establish a regime to sanction activities that damage marine ecosystems, destroy mangrove forest or cause water pollution. Indonesia passed its first environmental management legislation in 1982, with Law No. 4/1982 on Basic Provisions on Environmental Management. In 1997 this law was replaced by Law No. 23/1997 on Environmental Management. Despite this long history, to date there have been few successful cases of law enforcement through the courts.[6]

Almost immediately after the promulgation of Law No. 23/1997, commentators began to criticize its provisions.[7] It has now been reviewed and a new draft law has been submitted to the national parliament for consideration. One of the most persistent criticisms has been the law's failure to support an 'integrated approach' to environmental management that would allow related issues to be tackled together, building on cross-departmental and sectoral cooperation, engagement with all stakeholders and the integration of regional and national policies. Poor integration is said to be a result of existing institutional arrangements and the failure to link environmental and natural resource law (Rahmadi 2006: 128–31).

Under the existing institutional arrangements, implementation of environmental law and natural resource management relies on the initiative of the relevant sectoral ministry (Rahmadi 2006: 130). The difficulty is that the State Ministry for the Environment, as a non-departmental ministry, has little authority other than to issue standards and guidelines and coordinate the activities of the sectoral ministries. Sectoral ministries such as the Ministry of Marine Affairs and Fisheries, Ministry of Forestry, Ministry of Culture and Tourism, Ministry of Energy and Min-

---

6 From 2002 to 2007 there were 24 prosecutions for water pollution and toxic waste, resulting in three convictions, four acquittals and 17 bonds. One prosecution for environmental damage resulted in an acquittal. Of seven forest fire prosecutions, there was one conviction, five acquittals and one bond. Figures provided by Ilyas Asaad, Deputy Minister for Environmental Compliance, 11 August 2008.

7 Faure and Niessen (2006) provide a recent and comprehensive assessment. See also Koeswadji (1993), Santosa and Fjellstrom (1997), Indonesian Centre for Environmental Law (1999), Koesnadi (1999), Rangkuti (2000), Nicholson (2001) and Silalahi (2001).

erals and Ministry of Industry — all relevant for the protection of coastal and marine resources — are responsible for regulating the activities that fall within their jurisdictions, including issuing utilization/exploitation permits (*izin pemanfaatan*) for fish-farming, tourism, mining, forestry, industrial and other environmentally sensitive ventures. Each ministry is expected to implement its own laws to prevent environmental degradation, for example with regard to clean-up activities or restoration of the environment when damage occurs (Rahmadi 2006: 131). In general, the environmental performance of the ministries has been disappointing, particularly in relation to carrying out and implementing environmental impact assessments (Rahmadi 2006: 131).

Environmental agencies are often reluctant to enforce the criminal provisions of environmental law unless administrative sanctions have first been imposed by the relevant sector. Administrative sanctions are imposed where there is a failure to obtain a licence or a breach of a licence condition. But in environmental law, it is often the case that sectoral licences do not contain conditions relating to environmental protection or management that could become the basis for administrative sanctions. Also, if enforcement is seen to be 'administratively dependent', it can result in a situation where grave environmental damage is not prosecuted directly, as it is expected that it must first be shown that the action was taken without a licence or in breach of a licence condition (Faure 2006: 192).

Many of Indonesia's environmental law commentators have called for a ministry with strong enforcement powers in both administrative and criminal law, but this would require a major change to current institutional arrangements. It has also been proposed that Indonesia should have a detailed environmental code to integrate the different statutory regimes and government regulations. To date, the existence of overlapping jurisdictions has made for confusion, and at times contradiction, between the laws prepared by various sectors. However, as Rahmadi (2006: 136) observes, the idea of a detailed code does not fit with the conventional paradigm of law making in Indonesia whereby statutes are usually brief and general.[8]

An obvious gap in the law is that there is no simple provision to the effect that a criminal prosecution can be based on the failure to obtain a licence or to comply with a licence condition (Stoink 2006: 186). However, the drafting of legislation is also at fault. One author reached the

---

8  In the author's experience, Indonesian legislative drafters are frequently surprised that a statute can be drafted as a complete legal instrument. This points to the need for more emphasis on comparative law in both legal education and professional training.

conclusion that the sanction provisions were so poorly structured and contained so many complex conditions that they were very difficult to apply effectively in legal practice (Faure 2006: 209).

I do not have the space to discuss every aspect of the criminal sanctions provisions, but I would observe that the necessity to prove fault (intention or negligence) in all offences imposes a heavy evidential burden on prosecutors. A legitimate topic for discussion in Indonesia would be whether fault liability should be retained for all offences in environmental law. A trend in some countries has been to impose absolute or strict liability in relation to offences where there is considered to be an overriding public interest to protect human health, safety or the environment.[9] Proof that an offence has occurred then becomes relatively simple, and it falls to the judge to exercise judicial discretion in sentencing after considering all aspects of the case. Strict liability has been a significant deterrent and is particularly helpful in the prosecution of corporations, where imprisonment is not practicable and where it is frequently difficult to obtain evidence of actual intention or negligence.[10]

## THE FISHERIES LAW: TOWARDS AN ECOSYSTEM APPROACH?

### Legislative support for an ecosystem approach to fisheries

In the past, the main objective of fisheries management was to provide support for the exploitation of fishery resources by Indonesians, and especially by poor fishermen. There was little concern about overfishing or the possibility of the wholesale destruction of fish stocks. Internationally, fisheries management experts now focus more on the sustainable use of aquatic ecosystems, which demands a more sophisticated approach involving the careful assessment of whole ecological systems. The United Nations Food and Agriculture Organization (FAO) has been particularly active in promoting an ecosystem approach to fisheries.[11] However, it is acknowledged that there are numerous difficulties in implementing this approach. For example, the boundaries of an ecosystem, whether it is an estuary, a large bay, a coastal zone or a large marine ecosystem, will usually not match the limits of the relevant legal jurisdictions. Also, effective

---

9   Two examples from Australia are section 120 on water pollution in the Protection of the Environment Operations Act NSW and section 24A on offences relating to marine areas in the Environmental Protection and Biodiversity Conservation Act.

10   In Australia, to balance the overriding of traditional legal rights, there is a defence of honest and reasonable mistake of fact (Bates 2006: 240).

11   See http://www.fao.org/fishery/topic/2880/en, accessed 4 September 2008.

management at an ecosystem level is likely to require the cooperation of different authorities or an adjustment of administrative boundaries (Garcia et al. 2003).

Law No. 31/2004 on Fisheries, which came into effect in October 2004, was the first major piece of legislation to be passed after the creation of the Ministry of Marine Affairs and Fisheries in 1999. It gives the minister authority to determine protected fish species and protected areas, including national marine parks (article 7(5)). It also gives the minister responsibility for planning fisheries management, determining fish stocks and setting allowable catch levels. The law contains many new provisions on fish cultivation, fish processing, food additives, fishing enterprises and licensing, but many of the innovations are only alluded to and the detail will not be available until implementing regulations are issued.

The Fisheries Law strengthens the prospect of implementing an ecosystem approach to fisheries in Indonesia, stating that 'within the framework of managing fishery resources, ecosystem conservation, conservation of fish species and conservation of fish genetic resources are to be carried out' (article 13(1)). Further provisions on the conservation of ecosystems, fish species and fish genetic resources are to be provided in a government regulation (article 13(2)). The Centre for Fisheries Data and Information (Pusat Data dan Informasi, or Pusdatin) established under article 46 publishes an annual report on conditions in the Indonesian fisheries sector. The website of the Ministry of Marine Affairs and Fisheries, especially its Marine and Fisheries Statistical Information System (http://statistik.dkp.go.id/), is another important source of fisheries statistics.

### Expansion of marine reserves

The creation of marine reserves protected from potentially damaging human activity is a crucial component of an ecosystem approach to fisheries. Recent research confirms that marine reserves are crucial in protecting fish stocks against overfishing, and that no-take marine reserves, in which fishing is completely banned, can lead to a very rapid resurgence of commercially valuable fish species.[12] Especially significant is the conclusion of another study that coral reef marine protected areas established by local people for traditional uses can be far more effective at protecting fish and wildlife than reserves set up by governments expressly for con-

---

12 'Exploited fish make rapid comeback in world's largest no-take marine reserve network', *Science Daily*, 25 June 2008, available at http://www.sciencedaily.com/releases/2008/06/080623125012.htm, accessed 3 September 2008.

servation purposes.[13] Although large, permanent marine protected areas may provide the best protection for species that are at risk from overfishing, a combination of large reserves and traditionally managed systems is likely to represent the best overall solution to meet conservation and community goals and reverse the degradation of reef ecosystems.

An approach that accords with these research findings is developing in Indonesia within the framework of regional autonomy. Ministerial Regulation No. 17/2008 on Conservation Areas in Coastal and Small Island Areas, passed in 2008, deals with the categorization, determination and management of conservation areas. In June 2007, Indonesia had seven marine national parks (*taman nasional laut*), 18 marine tourism parks (*taman wisata alam laut*), seven marine protected areas (*suaka margasatwa laut*), nine marine nature reserves (*cagar alam laut*), 13 regional marine conservation areas (*kawasan konservasi laut daerah*), 15 proposed regional marine conservation areas (*calon kawasan konservasi laut daerah*), 28 regional marine protected areas/regional mangrove protected areas (*daerah perlindungan laut/daerah perlindungan mangrove*) and 10 fish sanctuaries (*suaka perikanan*) (MMAF 2006: Table 1.2). The Ministry of Marine Affairs and Fisheries intends to expand marine protected areas to 10 million hectares by 2010 and 20 million hectares by 2020. The ministry's statistics show that by June 2007, the national government had set aside almost 7 million hectares.

Alternative systems of management are being tested within the framework of national park zoning plans and at the regional level through community-based management. The experience of national parks on land shows that the creation of large-scale protected areas has relatively little value if it is not accompanied by effective management. Strategies need to be tailored to local social, cultural and economic conditions and to be based on management by local communities themselves.[14] The effectiveness of strategies needs to be monitored and evaluated so that they can be adapted to needs as they arise. While the research shows that marine reserves can work, they need to be policed effectively and supported by effective law enforcement. In line with regional autonomy and a community-based approach, facilities are being handed over to coastal communities (MMAF 2006: Table 5.2). There were 759 community control groups (*kelompok masyarakat pengawas*) in 2006 and the number is continuing to expand (MMAF 2006: Table 3.3).

---

13  'Marine protected areas: it takes a village, study says', *Science Daily*, 28 July 2006, available at http://www.sciencedaily.com/releases/2006/07/0607271 80504.htm, accessed 3 September 2008.

14  One long-term program of this type is the Coral Reef Rehabilitation and Management Program (Coremap); see http://www.coremap.or.id/.

## LAW ENFORCEMENT IN THE FISHERIES LAW

### Obligations and prohibitions

The fundamental legal building blocks for effective law enforcement are clearly drafted obligations and prohibitions backed by proportionate sanctions. The Fisheries Law partially succeeds in this regard. An obligation has been imposed on 'any person who carries out a business or activity in fisheries management' to comply with the determinations of the minister (article 7(2)). Fisheries businesses would therefore have to comply with the minister's decisions on allowable catch levels, types of fishing gear, vessel monitoring, marine reserves, protected species and so on (article 7(1)). However, definitional issues could potentially arise in enforcement due to the lack of a definition of 'business or activity' as well as the failure to relate these obligations to artisanal fisheries.

It should be noted that these obligations have been designed to be enforceable in criminal law. Unusually, the Fisheries Law states that any person who breaches the provisions of article 7(2) may be liable to a fine of up to Rp 250 million (article 100). This has been the subject of some comment. It has been argued that under Indonesian law, article 7(2) can provide the basis only for an administrative sanction (such as a revocation of a licence) and not a criminal sanction.[15] The sanction relies on a determination of the minister being breached, and a ministerial determination does not normally provide the basis for a criminal sanction in Indonesian law.

The drafting of prohibitions is less problematic for effective enforcement. The law provides a detailed list of prohibitions, including one against owning, controlling, bringing and/or using (a) fishing gear and/or the means to catch fish that does not comply with the required size; (b) fishing gear that does not comply with conditions or standards; and (c) prohibited fishing gear (article 9). More general prohibitions are provided against taking action that may cause pollution and/or damage to fish resources and/or their environment (article 12(1)). Other prohibitions applicable to fish cultivation have also been formulated (article 12(2–4)).

A key provision is the prohibition against 'fishing (*melakukan penangkapan ikan*) or fish cultivation that uses chemical substances, biological substances, explosives, gear and/or methods and/or structures that are

---

15 'DPR kecam "pasal banci" UU Perikanan No. 31 Hadapi "Illegal Fishing"' [House of Representatives condemns 'transvestite article' in Fisheries Law No. 31 on 'Illegal Fishing'], *Antara News*, 6 July 2008, available at http://www.antara.co.id/arc/2008/7/6/dpr-kecam-pasal-banci-uu-perikanan-no-31-hadapi-illegal-fishing/, accessed 13 September 2008.

able to damage and/or endanger the preservation of fish resources and/ or the environment within the fisheries management territory of Indonesia.[16] This prohibition applies to 'any person' but is, in addition, specifically imposed on the shipmaster or captain of a ship, the owner of a ship, the owner of a fishing enterprise, the person responsible for a fishing enterprise, and the operator of a fishing enterprise or fish cultivation enterprise (article 8(1–4)). This provision has proven difficult to enforce where it is interpreted as requiring evidence that fish have actually been taken using the prohibited method.[17] Given the way it has been drafted, a court may hold that it is not sufficient for a prosecutor to adduce evidence that a ship had explosives or biological substances on board, without eyewitness evidence of their actual use.

## Oversight and sanctions

An essential aspect of enforcement is effective oversight of fishing activities. The Fisheries Law is very scant in detail as to who is to do what in oversight and simply states that enforcement is to be the subject of a government regulation. It establishes that oversight is to be carried out by fisheries supervisors (*pengawas perikanan*), to consist of both investigative and non-investigative officers (article 66(1–3)). It states that the community may contribute to fisheries oversight but provides no details apart from the suggestion that the community may report a suspected violation of the Fisheries Law (article 67, elucidation).

The central government is to provide the infrastructure for oversight, including boats, a monitoring system and piers. Overseers are to be equipped with firearms and other protection (article 69), with further details to be provided in a government regulation. Aji Sularso, the Director-General of Supervision and Protection of Marine and Fishery Resources, has stated that to cope with illegal fishing within its EEZ, Indonesia needs about 79 large ships but has only 21 medium-sized ships. Cooperative arrangements are being worked out with Australia and the Philippines to allow for joint patrols in adjoining waters.[18]

The new law contains particularly severe penalty provisions. For example, the maximum penalties for the intentional use of chemical or biological substances or explosives are:

---

16  Arguably, this would have been covered by the previous law as well.
17  Comment made by Rili Djohani at the Indonesia Update conference, Australian National University, Canberra, September 2008.
18  'Illegal fishing still rampant in Indonesia amid intensive operations at sea', *Antara News*, 7 August 2008.

- six years imprisonment and a fine of Rp 1.2 billion for 'any person';
- 10 years imprisonment and a fine of Rp 1.2 billion for a ship's captain; and
- 10 years imprisonment and a fine of Rp 2 billion for the owner of the fishing vessel.

Any person who intentionally violates the provisions on fishing gear in article 9 may receive a maximum of five years imprisonment and a fine of Rp 2 billion (article 85). Heavy sanctions are also provided for intentional failure to obtain the necessary licences.

It would seem that, depending on the severity of the offence, the full weight of these sanctions would not always be necessary, thus opening up a wide area for the exercise of judicial discretion in sentencing. More detailed provisions specifying a range of offences, from less to more serious, and the accompanying sanctions, would help to create greater legal certainty regarding the sentences that are likely to flow from the breach of a particular provision.

### Fisheries cases

Statistics on enforcement action are available on the Ministry of Marine Affairs and Fisheries' website.[19] Data on the number of fisheries violations per province (for 2003–07), the number of violations that have been processed by type of violation (for 2003–07) and the status or outcome of the prosecution process in each province (for 2007) are now readily available. These data are not complete, as can be seen from the absence of figures on violations in Papua during 2007 and for some other provinces in recent years. In Papua alone, it has been reported that 28 cases of illegal fishing had been filed for prosecution and many more were being processed as of April 2008.[20] Clearly it is logistically difficult to collate all the data, especially on smaller violations. For example, the author is aware of at least five cases that had been prepared for prosecution by the Department of Fisheries in West Nusa Tenggara, but only two are recorded in the data for 2007.[21] At the present time a proper analysis of trends cannot be made. However, online public access to information is

---

19 See http://statistik.dkp.go.id/download/StatistikKP_2007/Buku_3_3.htm.
20 'Sorong: illegal fishing marak di Papua selatan' [Illegal fishing popular in south Papua], 27 March 2008, available at http://konservasipapua.blogspot.com/2008/03/sorong-illegal-fishing-marak-di-papua, accessed 5 September 2008.
21 Copies of the case files were provided to the author by Dr Gatot Wibowo, Professor of Law, University of Mataram. The details are as follows.

a major step forward in creating greater transparency and accountability in law enforcement.

The official website states that 58 decisions were handed down across Indonesia in 2007. A final decision was made in 35 cases but the website does not provide any information on sentencing. In a decision that has been hailed as 'clear and resolute' and 'capable of producing the desired deterrence effect',[22] in April 2005 the district court of Sungailiat, in Bangka district in the province of Bangka-Belitung, imposed heavy sentences on four crew members alleged to have used prohibited substances or fishing gear. Each of the defendants was sentenced to four years gaol and fined Rp 75 million; in addition, the two foreign fishing vessels involved in the case were seized.

An important innovation is the establishment of a Fisheries Court within the National Civil Courts of General Jurisdiction (article 71(1–2)). The court in Medan has been operational since October 2007 and additional courts will be established within the district courts of North Jakarta, Medan, Pontianak, Bitung and Tual (article 71(3)). As it is difficult to obtain copies of judgments from district courts, analysis of Fisheries Law decisions has not yet developed to any extent. Notably, in the statistics on the processing of violations, the outcome is 'not known' in 21 cases where a judgment has been handed down. This highlights the problem of obtaining access to court decisions in Indonesia. The impor-

---

1 Sumbawa: Prosecution of seven fishermen for an intentional use of potassium to capture enough decorative fish to fill 50 plastic bags on 26 September 2006, in violation of articles 8(1) and 84(1) of the Fisheries Law.

2 Sumbawa: Prosecution of two fishermen for an intentional use of Akodan pesticide to capture fish to fill a small dam on 9 October 2006, in violation of articles 8(1) and 84(1) of the Fisheries Law.

3 Dompu: Prosecution of a fisherman for intentionally catching 10 turtles using a net on 31 August 2005 without a licence, in violation of articles 26(1) and 92 of the Fisheries Law.

4 Dompu: Prosecution of a fisherman for intentionally catching three turtles using a net on 9 September 2005 without a licence, in violation of articles 26(1) and 92 of the Fisheries Law.

5 West Lombok: Prosecution of five defendants for taking soft coral on 26 April 2008 using a mechanical hammer and compressor. This work had been going on for five months without a licence. The soft coral was sold to a company. The defendants were also prosecuted for using potassium to catch fish. Charges were brought for actions able to cause water pollution and/or damage to fish resources and/or their environment in violation of articles 8(1), 84(1) and/or 12(1), and 86(1) of the Fisheries Law.

22 'Keputusan pengadilan kasus "illegal fishing" harus jadi yurisprudensi' [Decisions in 'illegal fishing' cases must become jurisprudence], *Sinar Harapan*, 21 April 2005, available at http://www.sinarharapan.co.id/berita/0504/21/eko05, accessed 8 September 2008.

tance of making access to judgments more readily available cannot be overstated, especially as many of the decisions on fisheries cases would be handed down in remote areas. Tracking the outcome of these cases for a comparative study would be very valuable. Decisions of the the Constitutional Court, the Supreme Court and the Religious High Court are now freely available online.[23] This development has the potential to transform the transparency of judicial decision making in Indonesia and increase the scope for more vigorous scrutiny of judicial decisions.

## THE COASTAL AND SMALL ISLAND MANAGEMENT LAW

### The start to coastal management: defining the coastal area

The second law passed by the Ministry of Marine Affairs and Fisheries to protect marine resources was Law No. 27/2007 on the Management of Coastal Areas and Small Islands. Like the Fisheries Law, the Law on Coastal and Small Island Management only provides a framework, relying on many new implementing regulations before it can be fully implemented.

A complex issue has been how to define a 'coastal area'. This topic was debated extensively in the drafting of the new law, leading to the definition of a coastal area (*wilayah pesisir*) as 'the region of transition between terrestrial and marine ecosystems that is influenced by changes in the land and sea' (article 1(2)). Under this definition, to the extent that a stretch of water is influenced by changes on land, through either human or natural causes, it will be part of the coastal area. The definition allows for a management-based approach founded on actual human use of the sea and it also accommodates changes that occur in coastal areas over time due to social, economic and ecological influences. Thus, in theory, a coastal area may be narrow or broad, taking into account matters such as erosion/accretion and coastal activities such as tourism and fishing. On a coastline dominated by fishing, a much wider area of the sea would be included in the coastal area; on a relatively unused coastline, shelving into deep water, it might terminate close to the shore (French 1997: 192).

---

23  Decisions of the Constitutional Court are available at http://www. mahkamahkonstitusi.go.id/putusan_sidang.php. Decisions of the Supreme Court are available at http://www.putusan.net/app-mari/putusan/; 10,280 decisions were listed on the site in March 2009. Decisions of the Religious High Court are available at www.badilag.net. Indonesian legal information is also available at www.asianlii.org, a free, Asia-wide website run by the Australasian Legal Information Institute.

This definition, which is flexible in application, is qualified by a provision that a coastal area may extend no more than 12 nautical miles out to sea from the shore (article 2). Law No. 32/2004 on Regional Government and Government Regulation No. 38/2004 on the Division of Governmental Affairs between Provincial and District Governments state that within this 12-nautical-mile zone, the district government has authority over one-third of the sea, with the provincial government responsible for the remainder and for overall coordination. If an offence takes place within four nautical miles of the coastline, it will fall to the district government to take enforcement action, with the provincial government playing a coordinating role. If an offence occurs beyond the four-nautical-mile limit, the provincial government will have direct responsibility for enforcement. This means that the two levels of government must coordinate closely, firstly in determining the extent of the coastal area, and secondly in deciding exactly which level of government is responsible in any given situation.

### The context of integrated coastal management

The Law on Coastal and Small Island Management was drafted to take account of the principles underlying integrated coastal management.[24] A primary goal of integrated coastal management is to provide mechanisms to overcome sectoral and intergovernmental fragmentation. As stated by the FAO:

> Institutional mechanisms for effective coordination among various sectors active in the coastal zone and between the various levels of government operating in the coastal zone are fundamental to the strengthening and rationalization of the coastal management process. From the variety of available options, the coordination and harmonization mechanism must be tailored to fit the unique aspects of each particular national government setting.[25]

Therefore, the new law should be assessed according to how well it sets up coordination and harmonization mechanisms both horizontally and vertically.

At the national level, a presidential decree has been passed to form an Indonesian Maritime Council.[26] This consultative forum is to be chaired

---

24  Integrated coastal management is mentioned in the elucidation to the law (I(3): 'Scope'), and is reflected in the new planning system, particularly the zoning plan.
25  'Integrated coastal area management and forestry', http://www.fao.org/forestry/4302/en/, accessed 3 March 2009. See also Cicin-Sain, Knecht and Fisk (1995).
26  Presidential Decree No. 21/2007 on the Indonesian Maritime Council.

by the Minister of Marine Affairs and Fisheries, with all relevant sectors and stakeholders as well as business, academia and non-government organizations to be represented. At the regional level, the Law on Coastal and Small Island Management provides for a new set of planning instruments comprising a Coastal and Small Island Strategic Plan, a Coastal and Small Island Zoning Plan, a Coastal and Small Island Management Plan and a Coastal and Small Island Management Action Plan (article 7(1)).

Each regional government will be responsible for preparing a set of plans suited to its region (article 7(3)) and must involve the community in the process (article 7(4)). While there is specific mention of public participation and involvement of the business community (article 14(1–3)), the law does not set out any express obligations or procedures to achieve this. The provisions on vertical coordination are more explicit but still lack detail (article 14 (4–7)). The preparation of these plans is an onerous task and will require considerable assistance to regional government.[27]

Ministerial Regulation No. 16/2008 on Planning the Management of Coastal and Small Island Areas provides guidance on the preparation of planning instruments and an outline of the content of each plan. Space does not permit an examination of this instrument except to note that it contains quite detailed procedural provisions on the preparation of the plans, the factors that must be considered in their preparation, the relationship between district and provincial-level plans, zoning categories and technical aspects concerning mapping.

### Establishment of new rights to use marine resources

The Law on Coastal and Small Island Management establishes a new property right: the coastal water use right (*hak pengusahaan perairan pesisir*, or HP-3). It resembles the forestry use right (*hak pengusahaan hutan*, or HPH) and other similar rights in Indonesia. The HP-3 gives users the right to exploit an area stretching from the air immediately above the water down through the water column to the seabed (article 16(2)). It would also seem to grant the right to commercially exploit any small island located within that zone; however, these provisions are quite rudimentary (see articles 23–26) and will be elaborated in a ministerial regulation (article 26). It is anticipated that the HP-3 will allow users to carry out activities such as aquaculture (shrimp, fish, pearls), seagrass harvesting and eco-tourism (coral reefs).

---

27  This was the subject of an Asian Development Bank project known as the Marine and Coastal Resources Management Program (MCRMP), which ended in 2008.

Although not stated explicitly in the law, responsibility for issuing an HP-3 seems to be governed by the regional autonomy arrangements (article 23(4)), so that district governments will assess applications for rights up to four nautical miles from the coastline and provincial governments will assess applications for rights between four and 12 nautical miles from the coastline. The HP-3 has a time span of 20 years and can be extended for another 20 years. It can be transferred and used as security for finance (article 20(1)). In considering an application for such a right, governments must give consideration to the preservation of coastal and small island ecosystems, the needs of traditional communities (*masyarakat adat*) and the national interest as well as the transit rights of foreign ships (article 17(2)). An HP-3 cannot be granted over a conservation zone, a fish sanctuary, a service route, a harbour or a public beach (article 22).

The procedure for obtaining, registering and cancelling an HP-3 is to be formulated in a government regulation (article 20(4)), which reportedly has been prioritized for completion.[28] However, the new law does set out the conditions that must be fulfilled in order to grant an HP-3. Technical conditions include compliance with the relevant zoning and management plans, public consultation, and consideration of alternative proposals when assessing environmental impacts (article 21(2)). Administrative conditions include preparation of a plan on coastal resource utilization that is appropriate for the carrying capacity of the ecosystem, and an oversight and reporting system (article 21(3)). Operational conditions include empowering local communities in the vicinity of the HP-3, acknowledging, respecting and protecting the rights of traditional and/or local communities, providing access to beachfronts and estuaries, and rehabilitating the environment should any damage occur (article 21(5)).

The HP-3 has proven controversial, with some environmental and community groups concerned that it will lead to overexploitation of coastal resources and the marginalization of local communities. Environmental groups have drawn parallels between the HP-3 and the abuses of the HPH that occur in forestry.[29] Poor coastal communities are unlikely

---

28 'DKP gandeng BPN susun hak pengusahaan perairan pesisir' [Ministry of Marine Affairs and Fisheries holding hands with National Land Agency in preparing the coastal waters use right], 15 February 2008, available at http://news.okezone.com/index.php/ReadStory/2008/02/15/1/83803/dkp-gandeng-bpn-susun-hak-pengusahaan-perairan-pesisir, accessed 10 September 2008.

29 'Walhi dan nelayan demo tolak HP3 kamis' [Walhi and fishermen demo refusing HP3], 6 March 2008, available at http://www.kapanlagi.com/h/0000216705.html, accessed 11 September 2008; 'Kontroversi hak pengusahaan pesisir' [Controversy over coastal water use right], 14 March 2008, available at http://els.bappenas.go.id/upload/kliping/Kontroversi%20hak.pdf, accessed 11 September 2008.

to have the capacity or resources to apply for an HP-3. Moreover, while traditional community groups (*masyarakat adat*) are theoretically able to obtain an HP-3 (article 18(2)), in practice the narrow definition of *masyarakat adat* is likely to exclude many of the less traditional communities from doing so.

## Sectoral licensing

A crucial aspect of the integration of coastal resources management is to control sectoral activities that are likely to have a negative effect on the coastal and marine environment. In environmental and natural resource law, licensing can be used as a compliance tool to control the way activities are carried out, to obtain information and to achieve a balance between the exploitation and conservation of natural resources. Licences (*izin pemanfaatan*) allowing such activities as mining, oil exploration and drilling, land reclamation, forestry, coastal development, industry, tourism and fish farming are very relevant to coastal management. These licences are issued by the relevant sector, and although there is legislative provision for the inclusion of conditions concerning environmental protection, this rarely happens. An environmental impact assessment, if required, would lead to the preparation of an environmental monitoring plan and an environmental management plan, which would be incorporated into the sectoral licence. However, Indonesia's record on requiring and paying attention to environmental impact assessments is poor; in addition, not all activities that affect the coastal and marine environment require an environmental impact assessment.

A key aspect of achieving integration is to strengthen the link between planning and licensing. The zoning plan of each region will set out permissible activities, non-permitted activities and activities that are permitted on obtaining a licence.[30] For example, if approval is sought to establish a hotel in a zone that does not permit hotels, then the licence application must be refused. Each zoning plan will be passed as a regional regulation (article 9(5)), which will give it legal status. However, it is not clear exactly what turns on this status: conceivably, it could enable the passing of a regional regulation to impose sanctions should there be a breach of a zoning provision.

The management plan of each region will cover such matters as management policy, administrative procedure for permitted and prohibited uses of resources, the scale of priorities, public consultation in determin-

---

30   A zoning plan is defined in article 1(14). District governments also have the option of preparing a 'detailed zoning plan' containing details on each zone and the kinds and quantities of licences that may be issued; see the definition of a detailed zoning plan in article 1(17).

ing management goals and granting licences, reporting mechanisms and human resources (article 12(1)). According to Ministerial Regulation No. 16/2008, the management plan will also contain recommendations on licensing and the HP-3 (article 38(1)). A major weakness is that the management plan will not have legal status; rather, it will be issued as a governor's regulation (*peraturan gubernur*) or as a mayor's regulation (*peraturan bupati/walikota*) (article 37(1)). Thus it does not provide a solid legal basis for law enforcement. With legal status, a management plan could lay down minimum conditions for all licences, additional conditions relevant to particular sectors, and specific practices or technologies that must be adopted to protect the environment. Without such a firm legal foundation, the plan will function only as a tool to encourage regional sectoral agencies to include environmental provisions in licences rather than as a strong integrating mechanism.

## CONCLUSION

In passing the Fisheries Law and the Law on Coastal and Small Island Management, Indonesia has made major strides in developing a legal framework for improved protection of its marine and coastal resources. Significant developments include wide-ranging ministerial authority combined with legal recognition of the need to exploit fishery resources in a sustainable way and of the benefits of integrated coastal management. Indonesia has expanded its system of marine reserves in line with ambitious targets, and the number of community control groups is continuing to grow. The public now has online access to a range of information on marine and coastal resources, including the success rate of enforcement actions. In coastal management, the government has provided a practical definition of a coastal area, a new property right for coastal waters use and a system of planning and management for coastal areas and small islands. A number of fisheries courts have opened, and if their decisions are made readily accessible, as is already the case with the the Constitutional Court, the Supreme Court and the Religious High Court, fisheries law could become a fully fledged jurisdiction.

On the other hand, it is clear that there is still a great deal to be done in law making. Many detailed implementing regulations are necessary to turn the vision outlined in the Fisheries Law and the Law on Coastal and Small Island Management into a practical reality.[31] The drafting of legis-

---

31  The Fisheries Law requires 15 government regulations, two presidential decrees and six ministerial regulations. The Law on Coastal and Small Island Management requires four government regulations, six presidential regulations, one regional regulation and 12 ministerial regulations. Only a

lative detail usually presents a far more difficult task for policy makers and legislative drafters alike than the statements of principle and broad legislative provisions found in Indonesian statutes. Observers frequently comment that Indonesia's laws are adequate but implementation is lacking. However, closer analysis of legislative provisions often reveals significant impediments to effective implementation due to a lack of detail, particularly with regard to the allocation of responsibility and procedural provisions.

Despite longstanding criticism of Indonesia's sector-based approach to environmental management, the new laws do not provide strong mechanisms to integrate environmental protection measures with the regulation of sectoral activities. The reluctance to give the Coastal and Small Island Management Plan formal legal status as an overarching coordinative mechanism shows how difficult it is to constrain sectoral authority for the sake of environmental protection. There also appears to be continuing difficulty in drafting sanctions provisions, particularly in distinguishing the distinct roles of administrative and criminal sanctions. In addition, there would be value in exploring the gains to be made from imposing strict liability for some criminal sanctions. Rather than drafting impressively heavy criminal sanctions that apply across the board, more streamlined provisions with penalties tailored to the nature of the offence would be more practical and more likely to provide legal certainty for all involved.

## ACKNOWLEDGMENTS

The author would like to thank John Butcher for his assistance in locating references on the depletion of fish stocks in Indonesian seas and his encouragement to write about this topic.

## REFERENCES

Arifin, Z. (2004), 'Local millenium ecosystem assessment: condition and trend of the Greater Jakarta Bay ecosystem', Ministry of Environment, Jakarta, available at www.millenniumassessment.org/documents_sga/Indonesia%20MA2004_Final.pdf.

Bates, G. (2006), *Environmental Law in Australia*, LexisNexis Butterworths, Sydney.

---

small number of these implementing regulations have been passed — in fact, to the writer's knowledge, only those mentioned in this chapter.

Butcher, J.G. (2004), *The Closing of the Frontier, A History of the Marine Fisheries of Southeast Asia c. 1850–2000*, Institute of Southeast Asian Studies, Singapore.

Butcher, J.G. (2005), 'Bringing the state into explanations of fisheries depletions in Indonesia', paper presented to the People and the Sea III: New Directions in Coastal and Maritime Studies conference, Amsterdam, 7–9 July.

Cicin-Sain, B., R.W. Knecht and G.W. Fisk (1995), 'Growth capacity for integrated coastal management since UNCED: an international perspective', *Ocean Coastal Management*, 29(1–3): 93–123.

Faure, M.G. (2006), 'Towards a new model of criminalization of environmental pollution: the case of Indonesia' in M.G. Faure and N. Niessen (eds), *Environmental Law in Development: Lessons from the Indonesian Experience*, Edward Elgar Publishing, Cheltenham, pp. 188–217.

Faure, M.G. and N. Niessen (eds) (2006), *Environmental Law in Development: Lessons from the Indonesian Experience*, Edward Elgar Publishing, Cheltenham.

Fegan, B. (1999), 'Field report on March 1999 socioeconomic research on snappers in Lombok, Sumbawa and West Timor, Indonesia', Project No. FIS/97/165, Australian Centre for International Agricultural Research, Canberra.

Fegan, B. (2003), 'Plundering the sea', *Inside Indonesia*, 73(January–March): 21–3, available at http://www.insideindonesia.org/content/view/339, accessed 28 February 2009.

French, P.W. (1997), *Coastal and Estuarine Management*, Routledge, London.

Garcia, S.M., A. Zerbi, C. Aliaume, T. Do Chi and G. Lasserre (2003), 'The ecosystem approach to fisheries: issues, terminology, principles, institutional foundations, implementation and outlook', FAO Fisheries Technical Paper No. 443, Food and Agriculture Organization, Rome.

Gillet, R. (1996), 'Marine fisheries resources and management in Indonesia with emphasis on the exclusive economic zone', TCP/INS/4553, paper presented to the Workshop on Strengthening Marine Resources Management in Indonesia, Directorate General of Fisheries in cooperation with the Food and Agriculture Organization of the United Nations, Jakarta, 23 April.

Indonesian Centre for Environmental Law (1999), *Lingkungan Hidup dan Sumber Daya Alam pasca Orde Baru* [The Environment and Natural Resources after the New Order], Jakarta.

Koesnadi, H. (1999), *Hukum Tata Lingkungan* [Environmental Law], seventh edition, Gadjah Mada University Press, Yogyakarta.

Koeswadji, H.H. (1993), *Hukum Pidana Lingkungan* [Environmental Criminal Law], Penerbit PT Citra Aditya Bakti, Bandung.

Krisnandhi, Sulaeman (1969), 'The economic development of Indonesia's sea fishing industry', *Bulletin of Indonesian Economic Studies*, 5(1): 49–72.

MMAF (Ministry of Marine Affairs and Fisheries) (2006), *Buku Statistik Kelautan dan Perikanan Tahun 2006* [Book of Statistics on Marine Affairs and Fisheries 2006], available at http://statistik.dkp.go.id/, accessed 3 March 2009.

Morgan, G.R. and D.J. Staples (2006), 'The history of industrial marine fisheries in Southeast Asia', Regional Office for Asia and the Pacific, Food and Agriculture Organization, Bangkok, available at http://www.fao.org/docrep/010/ag122e/ag122e00.htm, accessed 5 March 2008.

Naamin, N., C.P. Mathews and D. Monintja (1995), 'Studies of Indonesian tuna fisheries, part 1: interactions between coastal and offshore tuna fisheries in Manado and Bitung, North Sulawesi', FAO Corporate Document Repository, Status of Interactions of Pacific Tuna Fisheries in 1995, available at http://www.fao.org/docrep/003/W3628E/w3628e0n.htm, accessed 3 March 2009.

Nicholson, D. (2001), 'Environmental litigation in Indonesia', *Asian Pacific Journal of Environmental Law*, 6(1): 47–78.

Numberi, Freddy (2008), 'Opening speech to the 30th Asian Pacific Fishery Commission (APFIC) main session', 11 August, available at http://www.dkp. go.id/index.php/ind/news/309/, accessed 12 September 2008.

Palma, M.A. and M. Tsamenyi (2008), 'Case study on the impacts of illegal, unreported and unregulated (IUU) fishing in the Sulawesi Sea', paper prepared for the Asia Pacific Economic Cooperation Secretariat, April, available at http://www.apec.org/apec/publications/all_publications/fisheries_working.html, accessed February 2009.

Pet-Soede, L. and M. Erdmann (1998), 'An overview and comparison of destructive fishing practices in Indonesia', *SPC Live Reef Fish Information Bulletin*, 4: 28–36.

Rahmadi, T. (2006), 'Towards integrated environmental law: Indonesian experiences so far and expectations of a future environmental management law', in M.G. Faure and N. Niessen (eds), *Environmental Law in Development: Lessons from the Indonesian Experience*, Edward Elgar Publishing, Cheltenham, pp. 128–81.

Rangkuti, S.S. (2000), *Hukum Lingkungan dan Kebijaksanaan Lingkungan Nasional* [Environmental Law and National Environmental Policy], second edition, Airlangga University Press, Surabaya.

Santosa, M.A. and K. Fjellstrom (1997), 'The Indonesian environmental management law 1997', *Asian Pacific Journal of Environmental Law*, 2(3–4): 366–72.

Silalahi, M.D. (2001), *Hukum Lingkungan dalam Sistem Penegakan Hukum Lingkungan Indonesia* [Environmental Law and Law Enforcement in Indonesia], Penerbit Alumni, Bandung.

Sodik, D.M. (2007), 'Combating illegal, unreported and unregulated fishing in Indonesian waters: the need for fisheries legislative reform', PhD thesis, University of Wollongong, Wollongong, available at www.library.uow.edu.au/ adt-NWU/public/adt-NWU20080905.114951/index.html.

State Minister for the Environment (2006), *State of Environment Report*, Jakarta

State Minister for the Environment (2007), *State of Environment Report*, Jakarta.

Stoink, F. (2006), 'Supervision and enforcement in the law concerning environmental management, Law No. 23 of 1997', in M.G. Faure and N. Niessen (eds), *Environmental Law in Development: Lessons from the Indonesian Experience*, Edward Elgar Publishing, Cheltenham, pp. 182–7.

UNESCO (United Nations Educational, Scientific and Cultural Organization) (2000), 'Reducing megacity impacts on the coastal environment: alternative livelihoods and waste management in Jakarta and the Seribu Islands', Coastal Region and Small Island Papers 6, Paris, available at http://www. unesco.org/csi/pub/papers/mega.htm, accessed 3 March 2009.

Zamansyah, T. (2000), 'Traditional fishers: up against trawling', *Samudra*, April, available at http://www.icsf.net/icsf2006/uploads/publications/samudra/ pdf/english/issue_25/art01.pdf, accessed 1 March 2009.

# 12 LEGAL AND ILLEGAL INDONESIAN FISHING IN AUSTRALIAN WATERS

*James J. Fox*

I should properly begin my narrative in 1728. This was the year in which a Dutch East India Company officer in Kupang first reported seeing Bajau Laut fishermen, with a fleet of some 40 small boats, gathering trepang on the south coast of the island of Rote. The date thus marks the beginning of recorded trepang fishing in the waters to the south of Timor. Within a few years voyages to Ashmore Reef were a yearly occurrence, and within a few decades Makassan voyages to 'Marege' were an established fixture off Australia's northern coast.

These voyages to Australia and in particular to the reefs situated between the mainland of Australia and the islands of Indonesia — numerous, various and for the most part undocumented — have persisted to the present day.[1] They constitute some of the earliest, and still continuing, connections between Australia and Indonesia. Designated as 'traditional', these connections have never been properly explored or adequately understood by the policy makers who have shaped formal maritime relations between the two countries, yet they have influenced — and continue to influence — the contemporary conduct of the relationship. They thus provide a context for any discussion of the fashioning of the formal agreements between Australia and Indonesia that were initiated in the 1970s to define and distinguish legal from illegal Indonesian fishing in Australian waters.

---

1 The two best sources of documentation on these voyages are Campbell Mac-knight's study of Makassan trepangers (1976) and Natasha Stacey's study of Bajau fishermen (2007). In preparing the final draft of this chapter, I have bene-fited greatly from comments provided by Natasha Stacey on an earlier draft.

The early 1970s, it should be remembered, were a period of particular optimism and mutual goodwill in Australia–Indonesia relations. President Soeharto visited Australia in February 1972 and Prime Minister Gough Whitlam reciprocated with a visit to Indonesia in August 1974, where the two held important talks in Wonosobo. It was in this spirit of cooperation that various legal matters relating to Indonesian fishing were agreed upon.

Since this chapter deals with both legal and illegal fishing, it is essential to begin by defining the differences between these two activities in terms of the legislation and agreements between Indonesia and Australia that were initiated in the 1970s.

## THE LEGAL CONTEXT FOR INDONESIAN FISHING

### An outline of initial agreements

Seabed negotiations between Indonesia and Australia were begun in March 1970 and in just 15 months, in May 1971, seabed boundaries were agreed upon in the Arafura Sea and the eastern part of the Timor Sea. A maritime demarcation line between Indonesia and Australia was identified by reference to 13 points, thus defining a clear boundary along a stretch of 520 nautical miles. However, a gap was left in this line—the so-called Timor Gap—that has remained unresolved to this day.

Establishment of this border, which came into force on 8 November 1973, created concern in Australia about the activities of the Indonesian fishermen who regularly sailed beyond these limits. In 1968, when Australia established its 12-mile fishing zone for the exclusive use of fishing vessels licensed under Australian law, the Australian government recognized the existence of traditional Indonesian fishing in its waters and allowed such fishing under two provisions: (1) that 'operations were confined to a subsistence level': and (2) that 'operations were carried out in the Declared Fishing Zone and territorial sea adjacent to the Ashmore and Cartier Islands, Seringapatam Reef, Scott Reef, Adele Island and Browse Island' (DFAT n.d.).

Australia's concerns over traditional Indonesian fishing were raised during the talks between President Soeharto and Prime Minister Whitlam and it was decided to hold formal discussions on the issue. These discussions, held in Jakarta on 6–7 November 1974, resulted in the Memorandum of Understanding relating to the Operations of Indonesian Traditional Fishermen, or MOU74. This document of seven paragraphs gave 'traditional fishermen' permission to fish 'twelve miles seaward off the baseline' of five small reefs or islets in the Australian exclusive fishing zone (see Map 12.1). The specific locations were (1) Ashmore Reef,

*Map 12.1*   *Traditional fishing zones agreed in the 1974 MOU*

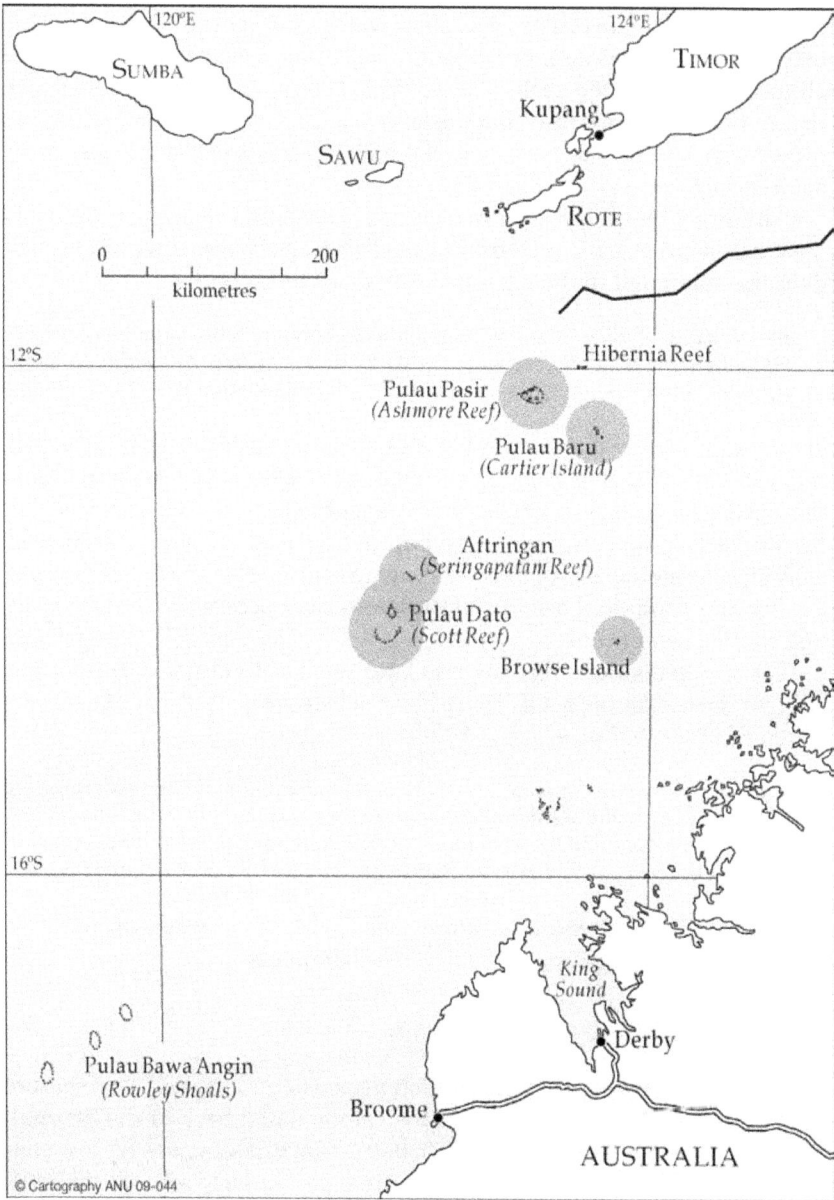

120°E

SUMBA

124°E

TIMOR

Kupang

SAWU

ROTE

0     200

kilometres

12°S

Hibernia Reef

Pulau Pasir
(*Ashmore Reef*)

Pulau Baru
(*Cartier Island*)

Aftringan
(*Seringapatam Reef*)

Pulau Dato
(*Scott Reef*)

Browse Island

16°S

King
Sound

Derby

Pulau Bawa Angin
(*Rowley Shoals*)

Broome

AUSTRALIA

© Cartography ANU 09-044

(2) Cartier Island, (3) Scott Reef, (4) Seringapatam Reef and (5) Browse Island.[2] The MOU made it clear that 'Indonesian fishermen will not be permitted to take turtles in Australian waters' but would be permitted to take 'trochus, bêche-de-mer [trepang], abalone, green snail, sponges and all molluscs' from the seabed adjacent to the five reefs and islets specified in the agreement. It did not establish a large sea area for fishing but rather a set of discrete locations in which fishing and marine gathering could occur.

Although the agreement recognized traditional fishing, it failed to designate who exactly such traditional fishermen were or would be. The defining statement in the agreement reads as follows:

> By 'traditional fishermen' is meant the fishermen who have traditionally taken fish and sedentary organisms in Australian waters by methods, which have been tradition over decades of time (MOU74, paragraph 1).

Defining 'traditional' by a double reference to 'tradition', specifically in relation to methods of fishing, left undefined who might be involved in this fishing. A history of at least 250 years of fishing in Australian waters by different Indonesian maritime populations—all of whom used basically similar methods—opened the way to multiple claims of 'traditional' fishing and prompted a series of problems over access that have continued to the present time.

The problems that were to arise later were not glimpsed at the time. A press release on the MOU issued by the Department of Foreign Affairs stated that:

> The talks were held in an atmosphere of friendly and mutual understanding of the problems of the Australian Government and also of the fact that fishermen from Indonesian villages have traditionally operated for many years in waters off the coast of Australia (News Release D24, 11 November 1974).

Based on the 1974 memorandum, the Australian Fisheries Act was amended to take account of the new arrangements.

### An outline of subsequent agreements

In the 1970s, Indonesian trepang fishing was concentrated at Ashmore Reef—a continuation of activities that can be dated back to the first half of the eighteenth century (Fox 1977). Bajau shark fishing was far less concentrated and occurred over a wider area, particularly around the reefs

---

2  Adele Island, which is much closer to the northwest coast of Australia than Browse Island, was dropped as a declared fishing zone in all subsequent discussions and determinations on traditional Indonesian fishing.

to the south of Ashmore and along the Kimberley coast (Stacey 2007: 91–5). There is so little specific documentation for this period that it is impossible to identify with certainty other possible fisher groups in these waters. In Indonesian as well as in Rotenese, Ashmore Reef is referred to as 'Sand Island' (Indonesian: Pulau Pasir, Rotenese: Nusa Solokaek; see Fox 1998). In English, the reef was named after Captain Samuel Ashmore, who sighted it in 1811.

From 1840 onwards, Ashmore was exploited by both the British and American ships that regularly visited the reef to gather guano to be used as fertilizer. Britain officially annexed the reef in 1878 and prohibited guano gathering in 1904 but it did not take formal possession of the reef until 1906. Britain transferred possession of the reef to Australia in 1933; this was recognized in the Commonwealth Ashmore and Cartier Islands Acceptance Act of 1933, which came into force in 1934. Western Australia was initially given administration of the islands but in 1938 this was transferred to the Northern Territory (Stacey 2007: 83–5). Following the agreements with Indonesia in the 1970s, the status of reefs within the Commonwealth of Australia required further legislation. With effect from 1 July 1978, the 1933 Commonwealth Ashmore and Cartier Islands Acceptance Act was amended to establish these islands as a separate Commonwealth territory.[3]

In 1979, the federal government extended the Australian fishing zone from 12 to 200 nautical miles; Indonesia followed suit by extending its fishing zone in March 1980. These mutual extensions created an overlap in national jurisdictions in the Timor Sea. This required the two countries to sign a further MOU, the Memorandum of Understanding on a Provisional Fisheries Surveillance and Enforcement Arrangement. Signed on 29 October 1981, it stipulated that neither country would take action against fishing vessels licensed by the other state beyond a demarcated 'provisional' surveillance and enforcement line. Although these arrangements did not affect traditional fishing, which continued as before, they did apply to motorized vessels seeking pelagic (or 'swimming') fish species. More importantly, seabed boundaries continued to apply for sedentary species on the sea floor, such as trepang.

A further step in the legal developments affecting relations between Indonesia and Australia came with the declaration, on 16 August 1983, of Ashmore Reef as a national nature reserve. This decision was taken for conservation reasons under the 1975 National Parks and Wildlife Conservation Act, prompted firstly by the Agreement between the Government of Japan and the Government of Australia for the Protection of Migra-

---

3   This act was later amended, with effect from 1 March 1988, to take into account the nature reserve status of Ashmore and Cartier.

tory Birds and Birds in Danger of Extinction and their Environment, which came into force on 31 April 1981, and secondly by the acceptance of an amendment to Article XI(3a) of the Convention on International Trade in Endangered Species of Wild Fauna and Flora (CITES), to which Australia and Indonesia were signatories. From the Australian perspective, this development required a revision to the arrangements regarding traditional fishermen. This involved several years of negotiations and a number of exchanges of documents.

Proposals to restrict fishermen's access to Ashmore Reef as a consequence of its official change in status were first made in August 1986 as part of an attempt to renegotiate the 1974 MOU. These were rejected by Indonesia as unacceptable. A subsequent proposal was put to the Indonesian Department of Foreign Affairs on 25 February 1988 setting forth Australia's position and the arrangements it intended to follow with regard to traditional fishermen from 1 March 1988. These arrangements still limited fishing to 12 nautical miles around the reefs and islands covered by the MOU. The changes proposed in this document were also judged unacceptable by Indonesian officials.

A breakthrough came at a meeting on 2 March 1989 between foreign ministers Ali Alatas and Gareth Evans, which allowed officials from both countries to meet on 28–29 April 1989 to work out new arrangements for traditional fishing. The minutes of these meetings contained 'Practical guidelines for implementing the 1974 MOU' (Environment Australia 2002: 70). They reaffirmed the earlier MOU74 agreement, placing renewed emphasis on traditional fishing methods and access by traditional sailing vessels. The first paragraph began with a statement that used 'tradition' or 'traditional' four times to make its point:

> Access to the MOU area would continue to be limited to Indonesian traditional fishermen using traditional methods and traditional vessels consistent with the tradition over decades of time, which does not include fishing methods or vessels using motors or engines (Environment Australia 2002: 70).

The guidelines contained two key changes. First, fishermen were prohibited from all fishing activities, including the gathering of sedentary species, in Ashmore Reef National Nature Reserve, though they were permitted to land at Ashmore's West Islet to replenish their fresh water supplies. Second, traditional fishing activities would no longer be confined to 12 nautical miles around each reef or isle, but would be allowed in an expanded area that has come to be known as the MOU Box—a defined block of sea territory between Indonesia and Australia that contained the principal reefs visited by Indonesian fishermen (see Map 12.2).

The minutes also stated that:

Map 12.2    *Traditional fishing zones agreed in 1989: the MOU Box*

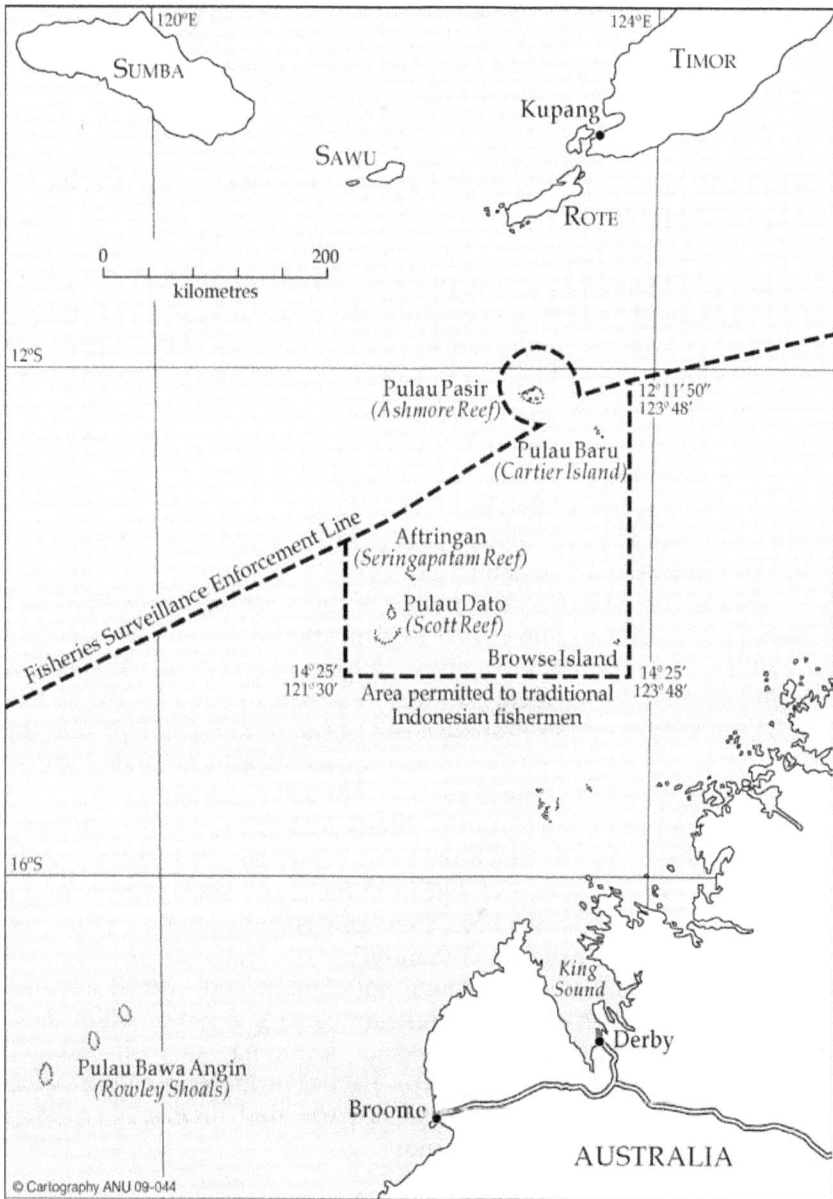

120°E

124°E

SUMBA

TIMOR

Kupang

SAWU

ROTE

0    200

kilometres

12°S

Pulau Pasir
(*Ashmore Reef*)

12° 11′ 50″
123° 48′

Pulau Baru
(*Cartier Island*)

Aftringan
(*Seringapatam Reef*)

Fisheries Surveillance Enforcement Line

Pulau Dato
(*Scott Reef*)

Browse Island

14° 25′
121° 30′    Area permitted to traditional
Indonesian fishermen

14° 25′
123° 48′

16°S

King
Sound

Derby

Pulau Bawa Angin
(*Rowley Shoals*)

Broome

AUSTRALIA

© Cartography ANU 09-044

The Indonesian and Australian Officials agreed to make arrangements for cooperation in developing alternative income projects in Eastern Indonesia for traditional fishermen traditionally engaged in fishing under the MOU. ... Both sides mutually decided to discuss the possibility of channeling Australian aid funds to such projects with appropriate authorities in their respective countries (Environment Australia 2002: 67, paragraph 8).

## THE FISHING COMMUNITIES OF THE SOUTHERN ISLANDS OF EASTERN INDONESIA

The fishing communities of eastern Indonesia are notable for their mobility and adaptability. This can be seen in the movement of many of these communities from the northern to the southern islands of the region over the past two centuries. Many of the most active fishing communities in the southern islands of eastern Indonesia — whether on Flores, Alor, Rote and Timor, or further east on Tanimbar, Aru and the southern coast of Papua — are comprised of migrants or the descendants of migrants from islands further to the north. The repositioning of these fishing communities has also had an effect on local communities, in some cases transforming their farmers into part-time fishermen.

The Bajau Laut are the most mobile of these fishing populations and they were the first to move their settlements southward. This migratory movement began in the eighteenth century and gathered speed in the nineteenth and twentieth centuries. The Bajau Laut have the largest number of settlements in East Nusa Tenggara, from Labuan Bajo to the Bay of Kupang. Bugis and Butonese settlements also expanded southward throughout the twentieth century (Fox 2000: 349–52).[4]

One of the interesting features of the southward expansion of Bugis, Buton and Bajau is that much of it has come, not from the island of Sulawesi as one might expect, but from the many small islands to the south and southeast of Sulawesi. These tiny islands, oriented to the sea as fishing and boat-building communities, did not have the capacity to continue to absorb their growing populations and instead adopted a strategy of exporting their surplus numbers. As a group, all of these islands — Passi Tallu, Sabalana, Bonerate, Karompa, Kayuadi, Kalatoa, Batuata, Wanci, Kaledupa, Binongko — have contributed significantly to the creation of new fishing communities, particularly in East Nusa Tenggara and the southern Maluku islands.

---

4   In the late 1990s, the Bugis were expelled from East Timor and were temporarily driven out of Kupang, while the Butonese retreated from settlements on Ambon. This may be seen as temporary, however; the Bugis settlement at Oesapa in the Bay of Kupang has largely been restored.

As a continuing historical process, this migration southward has tended to follow the same pattern. Young men leave in search of a livelihood, eventually establish themselves in some new settlement where others from the same island have previously settled, marry locally or, more often, return to marry a woman from their home village, and eventually bring her to the new settlement. Over time, links with the home island weaken and more marriages are contracted locally. In virtually all these settlements Islam provides a bond that facilitates the mixing of migrants from different islands.

The most prominent fishing settlements on the southern islands of eastern Indonesia are Oelaba and Pepela on the island of Rote; Namosain and Oesapa on the Bay of Kupang; Saumlaki on the southern tip of the island of Tanimbar; the port town of Dobo on the island of Aru; and Merauke on the Papuan coast. Each of these sites provides a harbour for the small-boat fishermen who have, until recently, regularly fished, either legally or illegally, in Australian waters.

While all of these settlements share common features, the differences between them are equally marked. The small fishing settlements of Oelaba and Pepela, for example, were initially established at different strategic locations on the island of Rote in the early years of the twentieth century, mainly by migrants of 'Butonese' origin.[5] Namosain was once the main port of the Rotenese in the Bay of Kupang. The coastal strip near the town of Kupang on which it is located was originally granted to the Rotenese by the Dutch East India Company in the seventeenth century. In the 1980s, by governor's decree, it was made the main port for all sailing vessels and has continued to expand since then as a near-suburb of Kupang. Communities from all the fishing populations of eastern Indonesia live there. Oesapa is located further to the east of the town of Kupang. It was previously a small Rotenese settlement, but since the 1970s it has expanded considerably through a large influx of Bugis migrants who engage in trade and fishing. Saumlaki is another small port, but one that is conveniently located for sailing into Australian waters and has thus attracted a mixed group of fishers from other islands. Dobo, by contrast, is a port town of historical importance since the nineteenth century. It was once a major port for the trade between Makassar and the Papuan coast. Although no longer as important as it once was, Dobo continues to provide a seasonal rendezvous point for fishermen from numerous small islands further to the east and, as a consequence, still attracts new settlers from these islands. Merauke is the premier southern fishing port of the province of Papua; its harbour is filled with both large and small vessels.

---

5   The term Butonese actually embraces several distinct ethno-linguistic groups. What unites them is a heritage as subjects of the former Sultan of Buton.

It represents a new frontier for the small-boat fishermen of eastern Indonesia, and in recent years fishermen who previously sailed from Dobo have begun to shift to Merauke.

The populations of these fishing settlements continue to expand at rates that reflect the opportunities they appear to offer. There is also migration, both on a seasonal and permanent basis, from nearby communities to the more prominent fishing settlements. Each settlement can be said to have a network of established connections linking it to other fishing communities in the region. In addition, the fishing communities often recruit temporary crew from neighbouring farming communities, which in some cases have themselves taken up voyaging into Australian waters. Investigating the history of any one of these settlements uncovers links across the region.[6]

Added to this mix of eastern Indonesian fishermen are Madurese from the tiny island of Raas and the even tinier island of Tonduk. Madurese sailors and traders have long played an important maritime role in eastern Indonesia. By one account, it was the Madurese who showed the Rotenese how to build their first long-voyage sailing boats (*perahu*). Madurese fishermen are involved in trepang gathering throughout Indonesia and have a tradition of fishing for trepang in Australian waters. The fishing fleet on Raas has now been modernized, leaving only about a dozen traditional *perahu* (or *lete-lete*) on Tonduk to carry on traditional sailing. All have engines fitted to them, which they remove and leave in Pepela before sailing south. This occurs on an annual basis; hence Raas and Tonduk must be added to the list of ports that regularly send *perahu* into Australian waters.[7]

---

6   In East Nusa Tenggara, for instance, Lamaholot-speaking populations from the island of Solor have a long tradition of maritime activities. The main rulers of Solor allied themselves with the Dutch in the first half of the seventeenth century and were given a coastal strip near the fort at Kupang when the Dutch established themselves there in 1652. Lamaholot speakers also settled on Pantar and around the Bay of Kalabahi on the island of Alor. Over time, Kalabahi Bay became an important trading point and settlement area for migrants from Sulawesi.

7   Australian documents identify three types of Indonesian fishing vessels. Type I refers to sailing *perahu*, also known in Indonesia as *lete-lete*. These are the boats with a lateen sail used by the Madurese fishermen from Raas and Tonduk. Type II refers to a large class of sailing *perahu* known as *lambo*. The *lambo* has a Western-style sailing rig originally developed from a Dutch model in the nineteenth century (Horridge 1979). *Lambo* are used by Rotenese and Bajau fishermen as well as most Butonese sailors. Type III refers to all motorized vessels, large or small, with or without the addition of a sail. (In Indonesia, it is common for sailing *perahu* to have auxiliary engines.) In the most recent Australian government documents, types I and II are lumped together, allowing a simple contrast between sailing and motorized vessels (see Stacey 2007: 100–1 for descriptions and drawings). The Australian

## THE TARGETS OF FISHING

Shark and trepang are the principal species targeted by Indonesian small-boat fishermen. Trochus and other shells are often gathered during trepang-gathering expeditions but shell gathering is rarely the main or exclusive focus of fishing efforts. Nevertheless the quantities of shells, particularly trochus, obtained on voyages in Australian waters are considerable.

Sharks are caught on longlines or, in some cases, with nets and their fins cut off.[8] This allows small boats to catch a large number of sharks but to reduce their take to a small quantity of fin. Gathering trepang involves more labour. Trepang can be picked up in shallow waters along reefs or gathered by diving in deeper waters. Most fishermen employ both methods. Once gathered, the trepang must be boiled on board the boats, salted and then sun-dried on deck before being bagged for storage below deck. This is usually a daily process and requires each trepang boat to carry a large supply of firewood. While nylon lines and hooks, although expensive, are all that is needed for shark-fishing expeditions, the outfitting of trepang boats, whose crews may spend a month or two gathering their catch, requires substantial quantities of rice and drinking water, salt, kerosene for lamps (since most gathering takes place at night) and at least a truckload of firewood.

Both trepang and shark fin are intended for sale to the Chinese market. Trepang has long been part of a well-established 'China trade'. Progressively, over a period of many centuries, different products from eastern Indonesia — first cloves, then sandalwood, then nutmeg, and finally trepang — were gathered into this trade. It was the search for trepang by so-called Makassans — actually a mix of seafarers, Makassarese, Bugis and Bajau — that opened the north coast of Australia to this trade.[9]

There is a developed market in Indonesia for both trepang and shark fin. Local agents handle these products for intermediary agents who generally sell them on to merchants in Surabaya, who are the main export suppliers to Singapore, Hong Kong and elsewhere.[10] For those at the top

National Maritime Museum has a *lete-lete* in its collection (Mellefont 1988, 1991); the Northern Territory Museum has a *lambo* in its collection (Stacey 1992).

8   Under Australian law, the practice of shark finning is prohibited. In the Australian industry, the whole shark must be returned to port.

9   Fox (2008) examines the role of trepang and shark fin in the China trade and how their pursuit eventually drew the northern coast of Australia into a wider Southeast Asian trade with China.

10  For centuries, Makassar was the great emporium for trepang. Alfred Russel Wallace provides a remarkable description of Makassar's trade in trepang in the middle of the nineteenth century: '"tripang" or sea-slug are obtained by

of this marketing chain, this is a lucrative market—or was when supplies were abundant and Indonesia was the world's leading source of both shark fin and trepang. Although reliable information on this trade is limited, the supply of both these valuable marine products from within Indonesia appears to have declined considerably over the past decade. Their increasing scarcity has raised the prices of these products, prompting poor fishermen to take greater risks to obtain them.

The market discriminates between different varieties of shark and trepang, setting a range of prices according to size and variety.[11] The higher-priced varieties are sought after and so, where possible, specifically targeted by fishermen. Prices rise as depletion advances. Thus, for example, hammerhead and sawfish sharks along with shovelnose rays command a higher price than black tip and sandbar sharks, which are among the most common varieties caught by Indonesian fishermen. According to the fishermen themselves, the numbers of high-value shark have fallen significantly. Whereas in the 1990s hammerhead sharks and shovelnose rays were regularly caught at the southern end of the MOU Box, from 2000 onwards fishermen say that these varieties are only to be found in or near the Gulf of Carpentaria.

In the case of trepang, the prickly redfish (*Thelenota ananas; nanas* or *nenas* in Indonesian) and white teatfish (*Holothuria fuscogilva; koro susu* in Indonesian) command three times the price of ordinary leopardfish or flowerfish sea-cucumbers (*Bohadschia argus* and *Pearsonothuria graeffei;* both called *bintik* in Indonesian) and are becoming rarer. Some varieties that were once gathered in substantial quantities at Scott Reef, such as stonefish (*Actinopyga lecanora; obor* in Indonesian) and amberfish (*T. anax; duyung* in Indonesian) are now becoming a memory.

## LEGAL FISHING IN THE MOU BOX

Since 1974, Indonesian fishermen have been permitted to enter Australian waters and fish legally on a regular annual basis. To fish legally, fish-

---

shiploads for the gastronomic enjoyment of the Chinese' (Wallace 1869: 158). In contemporary Indonesia, Surabaya has eclipsed Makassar in the trade of trepang and shark fin.

11  Precise identification of species is a complex task, because locally named 'varieties' of both shark and trepang do not coincide with taxonomic species. Names for varieties vary between localities. For an attempt to match fishermen's identifications with shark species, see Fox and Meekan (2006); for an attempt to identify trepang species, see Fox (2008). The examples I cite here are those where there is a reasonable fit between the Indonesian variety name and a recognized species.

ermen are required to confine their activities to the area defined by the MOU and use only 'traditional' methods of fishing and gathering. Until 1989, when the revision to the 1974 agreement extended the area open to legal fishing beyond the 12-nautical-mile limit of the reefs, most legal fishing involved the gathering of trepang and trochus and was concentrated at Ashmore Reef. The 1989 revision insisted on the use of sailing vessels only, ending the ambiguity surrounding the use of small motorized vessels to fish the various reefs.

When Ashmore was closed to Indonesian fishermen, trepang and trochus gathering shifted to other reefs, particularly Scott Reef, and shark fishing, which had always occurred, came to increasing prominence given the expanded sea area in which the fishermen were permitted to operate. Beginning in 1986, Australia stationed a boat at Ashmore for the main fishing months of the year to record arrivals.[12] Fishermen were expected to stop there to register their presence and receive an authorization stamp in their logbook—usually a simple exercise book of the kind used in schools.

### Developments in response to changing conditions

Diverse sources provide evidence of fishing by Indonesian *perahu* at Ashmore and later along the north coast of Australia from the first half of the eighteenth century (Macknight 1976; Fox 1977). In an oral narrative recorded in the nineteenth century, the Rotenese claim to have discovered Ashmore Reef during the early eighteenth century (Fox 1998: 104–10).[13] The Bajau Laut are another group for whom historical documentation of visits to Ashmore and other reefs exists (Stacey 2007). Complaints to the Commonwealth government by the Western Australian state government about Indonesian fishermen at Ashmore in 1923, which had to be referred to the British government, led to the transfer of Ashmore and Cartier to the Commonwealth of Australia in 1931 (Russell and Vail 1988: 14). A fishing survey conducted by the Commonwealth Scientific and Industrial Research Organisation (CSIRO) in 1949 aboard the *FRV Warreen* noted *perahu* at anchor at Ashmore and others in the vicinity at sea, and still more at Seringpatam Reef: an estimated 30 *perahu* loaded with dried fish, eel, shark fin, clam meat, trepang, turtle shell and

---

12  Originally, the Department of Arts, Sports, the Environment, Tourism and Territories hired a contractor to provide this boat and record vessels visiting Ashmore. This function was later taken over by Customs.
13  What gives some credence to this oral narrative is that three of the four rulers in the tale can be identified in Dutch East India Company records in the early eighteenth century.

considerable quantities of trochus. The fishermen had erected drying racks for clam meat and fish on Ashmore Reef (Russell and Vail 1988: 15). Only in 1974, when these activities were officially noticed by the Australian government, were some of them prohibited. The taking of turtle shell and clam meat, for example, was forbidden.

Little monitoring was done of Indonesian fishing activities until the 1970s. Western Australian Fisheries was the first authority to begin to board Indonesian *perahu* in the 1970s to establish their identity and create a record of their catch. Official Commonwealth records for Ashmore began only in 1986. In 1987, after Ashmore was declared a national nature reserve, the Australian National Parks and Wildlife Service commissioned the Northern Territory Museum to undertake a survey of marine resources 'to assess the impact of traditional fishing activities' at Ashmore. By the time it had completed its report, the survey team had available to it two and half years of records dating from April 1986 to June 1988. These showed that a total of 151 *perahu* had visited Ashmore during the period, some in successive years. The general conclusion of the survey was clear:

> Most of the regular visitors were from Roti. Rotinese boats predominated [in] the number of *perahu* visiting Ashmore Reef, and comprised one half to two thirds of the total number of vessels recorded in each year. Visits by these vessels constitute a regular and sustained fishing effort at Ashmore. Few other vessels were regular visitors, and the majority of *perahu* recorded (72.8 per cent) appeared in only one year (Russell and Vail 1988: 24).

Based on interviews conducted in September 1987, the authors reported that some of the *perahu* that made 'irregular' visits to Ashmore 'appeared to be fishing mainly on an exploratory basis' and 'were inexperienced at trepang and trochus fishing (Russell and Vail 1988: 24, 37). They also noted the presence of motorized *perahu* from Buton visiting Ashmore en route to other reefs, and speculated whether their numbers represented an increasing trend (Russell and Vail 1988: 37).[14] The authors observed what they described as a 'strong bimodal seasonal pattern' at Ashmore: Indonesian *perahu* would make their first visits in March and April and return for a longer season lasting from August to October. The same weather-dependent pattern continues to the present day, whereby *perahu* rely on prevailing winds and avoid sailing at those times of the year when the seas are rough and there are strong contrary winds.

---

14 It is possible that the number of motorized *perahu* prompted Australian officials, in the revision to the MOU in 1989, to insist on access by traditional sailing vessels only.

From interviews, it was apparent that the fishermen identified themselves according to their distinct activities:

> Most crew would either identify themselves as trepang, trochus, or shark fishermen. Trepang and trochus fishermen tended to collect almost anything of value while shark fishermen were less likely to be collecting other marine fauna. For trepang and trochus fishermen, the bulk of their catch would generally consist of trepang (Russell and Vail 1988: 41).

Perhaps the most revealing observations concerned the upswing in the Indonesian trepang market. Whereas the Madurese fishermen from Tonduk had long experience in trepang gathering at Ashmore, many others had shifted to more intensive gathering of trepang in the 1980s as prices increased and quantities of harvestable trochus declined.

Australia's attempts to ban fishing at Ashmore finally succeeded in 1989, when Australia and Indonesia reached agreement on the revision to the 1974 MOU. The closing of Ashmore to trepang fishing shifted fishermen's efforts to shark fishing within an enlarged MOU Box and directed trepang gathering to Cartier (briefly, until it was declared a marine reserve in 2000) and to Scott Reef and other reefs. However, the number of Indonesian boats continued to increase, including the number of *perahu* from Rote.

### Developments among fishermen on Rote in the 1990s

Changing conditions of access to Australian waters prompted the fishermen on Rote, who made up the majority of those sailing into the MOU area, to make a number of changes, not just in response to the new guidelines, but also in response to changing market conditions in Indonesia. The price of shark fin, for example, was increasing even more rapidly than that of trepang, offering fishermen who could position themselves to take advantage of the situation the possibility of greater profits.

Before 1989, there was a rough occupational division of labour among fishermen on Rote. The overwhelming majority of Rotenese fishermen were trepang and trochus gatherers, whereas the Bajau, though not exclusively shark fishers, made up a majority of those who fished for shark.[15] Beginning in 1989, many of these Bajau, particularly those from Mola on the island of Wanci, began to settle in Pepela, where they were given the coastal stretch known as Tanjung Pasir as a separate area of residence. Natasha Stacey has documented this movement of Bajau to

---

15  The list of names and ports of origin of the *perahu* visiting Ashmore between March 1986 and June 1988 (Russell and Vail 1988: Table 1) reveals a considerable number of vessels from known Bajau settlements on Wanci in the Tukang Besi Islands and Passi Tallu to the south of Selayar.

Rote in considerable detail, noting that it was accompanied by an influx of new traders who established their base in Pepela specifically to compete for the increasing trade in shark fin (Stacey 2007: 118–33). Pepela fishermen who had previously fished for trochus and trepang relied on the Bajau to join their crews and teach them shark-fishing skills.

Gradually through the 1990s, all of the *perahu* in Pepela shifted to shark fishing. At the same time, increasing numbers of Bajau from Wanci and Kaledupa moved to an ever more crowded Tanjung Pasir for the same purpose (Fox 1998: 127–9). Given the mobility of the Bajau, this happened quickly. In November 1992, there was a single Bajau house at Tanjung Pasir (see plate 6.2 in Stacey 2007: 128); by June of the following year, 113 Bajau families were living in Tanjung Pasir and more continued to arrive in each of the following years.

Developments in Pepela prompted changes elsewhere on Rote. The other main fishing settlement of Oelaba, located on the northwestern coast of Rote, had previously relied as much on small-scale interisland trade as on trepang gathering for its livelihood. The shift to shark fishing by Pepela fishermen opened a niche for Oelaba fishermen to increase their trepang-gathering activities in the MOU Box.

Poor returns from dryland farming combined with the possibility of employment on boats venturing into the MOU Box drew local Rotenese farmers into new sailing opportunities. Rotenese villagers from the settlements of Mae Oe and So'ao in Daiama, a village near Pepela, joined Pepela boats as crew and eventually added their own *perahu* to the shark-fishing fleet. At the same time, on the other side of the island, villagers from Hundi Huk and Dau Dulu joined the Oelaba *perahu* in trepang gathering. They, too, eventually developed their own small fleets. In need of more manpower, boat owners in Oelaba began recruiting men from the islands of Pura, Pantar, Treweng and Buaya near Alor.

A lengthy report by Fox and Sen (2002) has examined this period in some detail. Of the 1,678 voyage records in the Ashmore database covering the period 1988–99, 1,426 (85 per cent) identify vessels from the island of Rote. These records, which cover repeated voyages over a period of more than a decade, specifically identify, by name, 393 Rotenese *perahu*. These *perahu* constitute 93 per cent of all *perahu* listed by name and port in the database.[16] *Perahu* from Pepela accounted for 66 per cent of all voyages to Ashmore but 69 per cent of all vessels; *perahu* from Oelaba accounted for 16 per cent of voyages but 19 per cent of vessels. To these totals may be added voyages by *perahu* from Mae Oe and So'ao within

---

16  The database includes a significant number of records for which complete information, notably port of departure, was not recorded.

the Pepela network and *perahu* from Hundi Huk, Dau Dulu and Pantar in the Oelaba network (Fox and Sen 2002: 18–19). The other two identifiable groups of fishermen are the Madurese trepang fishers (mainly from Raas and Tonduk) who represent 5.6 per cent of voyages but 6.7 per cent of vessels and the Bajau shark fishers from Wanci and Kaledupa who identify themselves by their home island rather than their resettlement area of Tanjung Pasir. These fishermen represent 3 per cent of voyages but 5 per cent of vessels. Although the pattern of fishing activities changed in the 1990s and the number of fishermen increased, the Ashmore database reveals a significant continuity in the ethnic composition of 'traditional' fishermen. New recruits were mainly Rotenese farmers who took up fishing and a similar group of villagers from Pantar and Alor who joined the Oelaba fleet.

## ILLEGAL FISHING IN AUSTRALIAN WATERS

Just as legal fishing increased in the MOU Box during the 1990s, illegal fishing in Australian waters also increased. It was not confined simply to the area in and around the MOU Box but occurred across a broad front along the whole of the northern coast of Australia.

From 1988 to 1999, 48 sailing vessels were caught sailing beyond the eastern or southern limits of the MOU Box, half of them in search of shark, and some as far south as the Rowley Shoals and Kings Sound. A number were repeat offenders. During this period, 107 vessels were apprehended within the MOU Box, including 105 motorized vessels. (The other two were *perahu* caught trying to gather trepang at Ashmore.) The majority (75 per cent) were caught using diving equipment to fish for trepang. A large number of the boats were identified as coming from Sulawesi and most were apprehended in 1994; one fleet of 24 trepang boats was apprehended in September 1994 and another of 35 trepang boats was apprehended in November 1994.[17] For most of the period, there were only one or two apprehensions a year in the MOU Box, but in 1994 there were 63 apprehensions and in 1995 there were 21 (Fox and Sen 2002: 19).

In the early 1990s, a new wave of intrusions into Australian waters began with the arrival of small-boat shark fishermen sailing south from

---

17  Indonesian information indicates that most were part of what was known as the Sinjai fleet, a large contingent of boats that made their way around eastern Indonesia, fishing and gathering where they could. In 1994, without previous experience, they seem to have decided to venture into the MOU Box, relying on information from fishermen in Kupang and Pepela.

Dobo (Fox 1992). Between 12 and 15 March 1991, 29 *perahu* were appre-hended while fishing for shark. Although they had all sailed from Dobo, the crews and captains of most of these boats came from a variety of small islands in eastern Indonesia: Bonerate, Kalatoa and Karopa in South Sulawesi, Binongko in Southeast Sulawesi, Pomana and Wuring on Flores, and Binonko on Alor. All of these fishermen identified them-selves as Butonese and some had family connections with the fishermen on other boats: the Kalatoa and Karopa boats, for example, could be linked to those from Wuring. To the fishermen, the spectacular rise in the price of shark fin at the time was justification enough for venturing into Australian waters. A number of traders in Dobo were prominent in encouraging this marine trade (Fox 1992).

One recurrent refrain among the fishermen was that their own waters were being heavily overfished by larger vessels, often of foreign origin, impeding their attempts to pursue a livelihood. The number of sharks in Indonesian waters, by these fishermen's accounts, was diminishing.[18]

In the 1990s, the fishermen from Dobo sailing into Australian waters were joined by fishermen from Saumlaki and Merauke. Merauke, in par-ticular, expanded as a port for small-boat shark fishermen, many of whom began sailing deep into the Gulf of Carpentaria in search of shark.

Another development was the intrusion of much larger, more sophis-ticated motorized fishing vessels, called 'ice-boats' because they had the capacity to freeze their catch. These boats targeted several valuable spe-cies of fish: two species of red snapper (*Lutjanus malabaricus* and *Lut-janus erythropterus*; *ikan merah* in Indonesian); the gold-band snapper (*Pristipomoides multidens*; *godi* in Indonesian); and, to a lesser extent, the marble hawkfish (*Cirrhitus pinnulatus*; *kakap kecil* in Indonesian) (Fox and Meekan 2006).

The vessels involved in this fishery were based in a number of eastern Indonesian ports: Tenau near Kupang on Timor, Tual in the Kei Islands, Benjina on Aru Island, Merauke on the Papuan coast and, most signifi-cantly, Probolinggo on the north coast of East Java. The Probolinggo bot-tom longline fishing boats were originally based in Tanjung Balai on the Riau island of Karimun in Sumatra and initially fished in the South China Sea. In the 1980s and early 1990s, as operations in the South China Sea became more difficult, the fleet shifted its base to Java and its operations to the Arafura and Timor seas, using various local ports as substations. Ships in this fleet were equipped with global positioning systems (GPS) and radar and could make radio contact with one another. They would

---

18　For a discussion of illegal fishing in the Arafura Sea and its social effects, see Resosudarmo, Napitupulu and Campbell (2009).

position themselves along the border and enter Australian waters as the opportunity presented itself.[19]

### Developments in illegal fishing after 2001[20]

From 2001, there was a huge increase in illegal fishing in Australian waters from the key southern ports of eastern Indonesia: Pepela, Saumlaki, Dobo and Merauke. Some of the most radical developments occurred in Pepela, where purpose-built small boats known as *bodi* began a new wave of shark fishing. The smaller *bodi* had 24-horsepower engines and could accommodate three or at most four crew; they were heavily loaded with diesel fuel. The larger *bodi* had two even more powerful engines and could accommodate five to six crew members. Most of these *bodi* were constructed on islands further to the north and brought to Rote to replace Pepela's *lambo-perahu* fleet. Most carried a complement of five lines with 16 hooks per line, though some larger *bodi* carried six lines and even more hooks.

These boats were designed to make quick incursions into Australian waters. Initially most captains relied on GPS for precise navigation, but as sailing patterns became regularized, less use was made of this technology. Shorter incursions were directed to an area known as the 'Enterprise' or 'Operations' area (Perusahaan) because of the presence of oil rigs.[21] Here, according to fishermen, there was less Australian surveillance than elsewhere. Longer incursions went beyond Scott Reef at the southern end of the MOU Box to an area closer to the Australian coast known as Bawa Pulau Dato. Some boats ventured even further south into waters near the Rowley Shoals and closer yet to the Australian coast, a general area referred to as Bawa Angin or Masor.[22]

Shorter incursions would take three to five days; the longer and far more dangerous incursions lasted seven or eight days. For short trips, fishermen set themselves a target of seven shark; for longer trips, they set a target of at least 10 shark, generally the larger, higher-value ones. The key to the profitability of this fishing was a quick turnaround after each trip. Whereas a sailing voyage might take two weeks or more, it was possible to make several *bodi* voyages in a single month. By October 2005,

---

19  See Fegan (1999) for more detailed information on the Probolinggo boats.
20  This section of the paper is based on Fox et al. (2009), which contains more details on the development of illegal fishing during this period.
21  This area is roughly coincident with the lower half of what was previously referred to as the Timor Gap.
22  Bawa Angin was a traditional fishing ground for the Bajau Laut well before 2000 – probably dating back to a period before World War II (Natasha Stacey, personal communication).

Pepela's *bodi* fleet had increased to several hundred vessels and scores of shark-fishing vessels from Pepela were travelling into Australian waters on an almost daily basis. The main obstacle faced by the boat owners who controlled these operations was to obtain sufficient fuel to maintain the quick turnaround time between voyages, particularly after October, when a reduction in the fuel subsidy dealt a serious blow to the operations of *bodi* from Pepela.

While there was also an increase in the number of shark-fishing vessels from Saumlaki and Dobo, the biggest changes occurred in Merauke. Merauke provided a more strategically placed port for shark fishing, especially into the Gulf of Carpentaria, and thus attracted fishermen who had previously been based in Dobo or had used Dobo as a seasonal port. After a tsunami struck the north coast of Flores, fishermen from Wuring moved to Merauke, adding to the numbers available for shark fishing. As a rapidly expanding port, Merauke also drew a mix of fishermen from South and Southeast Sulawesi. It offered the advantage of proximity to a productive boat-building industry in Kumbe, approximately 60 kilometres from the town. By 2005, Kumbe was reported to be producing 30–50 boats of various sizes each year.

Whereas Pepela boat owners opted to reduce the size of their boats, Merauke boat owners opted for larger versions of the motorized vessels used in the area. A standard boat with a raised cabin superstructure would have five to seven crew members and carry roughly the same number of lines and hooks as most Pepela *bodi*. The newer Merauke boats, however, showed various innovations. They were larger and more powerful, and used double (or even triple) 23-horsepower in-board engines. They had a distinctive bow winch beam to combine longline and net fishing, with the winch beam needed to haul in the net. To use their nets, these boats needed to operate close inshore along the Australian coast.

The effect of these rapid changes in technology was an enormous build-up of fishing in Australian waters. By October 2005, it is estimated that there may have been as many as 300–400 *bodi* and other *perahu* operating out of Pepela, possibly 40–50 motorized boats out of Saumlaki, at least 100 out of Dobo and more than 150 out of Merauke. Sightings by Australian surveillance authorities jumped correspondingly. In September 2005, there were 1,272 sightings of Indonesian vessels in Australian waters. In response, over a period of two years, the Australian navy in cooperation with Customs carried out successive large-scale operations to deal with illegal fishing. The first of these was Operation Clearwater, conducted from 4 to 19 October 2005. It was followed by Operation Breakwater, which was carried out from 19 March to 4 April 2006.

The number of apprehensions of vessels soared. In October 2005, 59 vessels were apprehended – the highest number ever apprehended in a

Table 12.1   *Sightings and apprehensions of motorized vessels in Australia's northern waters*

| Month | 2005/06 | | 2006/07 | | 2007/08 | |
|---|---|---|---|---|---|---|
| | Sight-ings | Appre-hensions | Sight-ings | Appre-hensions | Sight-ings | Appre-hensions |
| September | 1,272 | 27 | 818 | 45 | 71 | 14 |
| October | 954 | 59 | 450 | 27 | 107 | 15 |
| November | 755 | 29 | 520 | 33 | 79 | 25 |
| December | 693 | 48 | 97 | 11 | 58 | 17 |
| January | 407 | 21 | 41 | 11 | 31 | 6 |
| February | 800 | 35 | 41 | 6 | 11 | 13 |
| March | 448 | 46 | 43 | 7 | 62 | 15 |
| April | 446 | 32 | 45 | 5 | 83 | 39 |
| May | 883 | 37 | 105 | 14 | 88 | 1 |
| June | 283 | 22 | 11 | 1 | 72 | 1 |
| July | 504 | 41 | 79 | 3 | 58 | 0 |
| August | 933 | 15 | 57 | 7 | 102 | 1 |
| Total | 8,378 | 412 | 2,307 | 170 | 822 | 147 |

*Source:* Australian Border Protection Command.

single month. Over the 12-month period to the end of August 2006, apprehensions reached 412 vessels. This high level of apprehensions served as a significant deterrent, with both sightings and apprehensions dropping over the next year. In the 12 months to the end of August 2007, 170 vessels were apprehended, and in the following year 147 (see Table 12.1).[23] Although far fewer than in 2005, apprehensions in 2008 were still high and indicative of a continuing problem.

## THE CONTINUATION OF LEGAL FISHING AT SCOTT REEF

During the period in which Pepela became a major port for illegal shark fishing, Oelaba and its satellite settlements of Hundi Huk and Dau Dulu continued to send boats to gather trepang legally at Scott Reef. They were joined each year by a small number of boats from Tonduk. This pattern

---

23   I wish to thank Tom Marshall, Director General, Border Protection Operations, for providing these data.

of sailing was set in place by the 1989 revision to the 1974 MOU that had closed Ashmore.

During the high season from August to October 2007, a total of 71 *perahu* visited Scott Reef, all gathering trepang and some also gathering trochus. Eleven of these *perahu* were from Tonduk and the rest from Oelaba, Hundi Huk and Dau Dulu. Many, perhaps a majority, of the *perahu* from Oelaba were manned by captains and crew from Pantar, Pura, Tereweng and Buaya in the district of Alor. Driving them to undertake these voyages was the rising price of trepang.

Given the number of fishermen who now regularly gather trepang at Scott Reef, the issue is one of the sustainability of the stock — a subject that was first raised at the time of the Makassan voyages to northern Australia and was again raised as the reason for the closure of Ashmore Reef and Cartier Island. The fishermen themselves are aware of the decline in trepang, but as long as the prices paid for it remain high, they will continue to concentrate on intensive gathering. This, then, is the dilemma they currently face — one that will, it is assumed, affect their future.

## CONCLUDING OBSERVATIONS

This chapter has focused on a 40-year maritime relationship between Australia and Indonesia and the particulars of Indonesian fishing during this period. Several observations can be made about this relationship with regard to both legal and illegal fishing.

First, in 1968 the Australian government's conception of 'traditional' fishing was based on a false assumption, one that has from time to time resurfaced. The stated assumption in the first documents was that traditional fishing operations were for subsistence purposes. In fact, insofar as they focused on two major products, trepang and shark fin, they were part of a centuries-old China trade. In contemporary terms, this means that this traditional fishing is attuned to a market, responds to changing prices and has been carried out to meet market demand. Over the past 40 years, market demand has risen enormously in response to growing affluence in the Chinese world.

Second, it is indicative of a certain continuity that many of the same 'ethnic' groups that historically participated in these specialized fishing efforts continue to be involved in fishing in Australian waters. Rote remains a critically important departure point for voyages to the south, as it has since the early part of the eighteenth century.

Third, since the time of the 1974 MOU, the notion of 'tradition' has thwarted discussion of access. The drafting of the memorandum is circuitous. The defining sentence of the first clause refers to 'tradition' three times without attempting to specify to whom it is meant to apply. Tradition is viewed more in terms of fishing methods than the historical practices of specific groups over time. The 1989 revision to the memorandum succeeded only in reinforcing this notion by insisting on access to the MOU by sailing boat.

The insistence on traditional sailing vessels officially and legally opened the MOU Box to a wide range of maritime populations in eastern Indonesia, many of whom had never previously been involved in fishing in Australian waters. The official access rulings resulted in what might be termed two distinct streams of fishermen accessing Australian waters, as observed by Russell and Vail in 1986. On the basis of a few years of data collected at Ashmore, they were able to distinguish between boats that visited the reef on an exploratory basis and those that came regularly on an annual or near-annual basis. In the 1980s, there was thus a recognizable core of boats that was continuing to conduct activities that, for the most part, had been going on from the time of the original memorandum or before.

Later analysis of the Ashmore database confirms these observations. Based on records of visits to the reef between 1986 and 1999, Fox and Sen (2002) found that vessels from Rote, Raas/Tonduk and Wanci/Kaledupa (the latter including Bajau fishermen who had settled in Tanjung Pasir on Rote) accounted for the vast majority (96 per cent) of the fishermen legally permitted to fish in Australian waters.

Almost another decade's records have since been added to the database. By all indications, they will confirm the same pattern. The Ashmore database is a log of boats that register at Ashmore before sailing further into Australian waters. Boats are registered by name of *perahu*, captain, owner and home port. It is therefore possible to use the database to trace the recurrence and succession of named vessels from specific ports, and even the succession of boat captains working for particular owners. Interestingly, in terms of continuity, some of the *perahu* identified in the 1986 Russell and Vail report could still be identified by name at Scott Reef in 2007.

The original memorandum referred to fishing in Australian waters 'over decades of time'. The Ashmore database represents four decades of 'traditional' fishing and would make it possible to identify with greater precision just who the 'traditional' fishermen are. As yet, however, this interrogation of the database has not been attempted.

Fourth, the exemption of 'traditional' Indonesian fishermen from the application of Australian laws on fishing and rules of management has been an unequivocal miscalculation that has dogged all subsequent discussions. Thus, for example, shark finning is prohibited under Australian law but is permitted to Indonesian fishermen in the MOU Box. Similarly, without limits on access or quotas on catch, Indonesian fishermen are officially able to deplete the resources they seek beyond recovery.

What began in 1974 as an agreement to allow so-called traditional fishermen, whose numbers at the time were unknown, to continue to sail to a few reefs in Australian waters has taken on a treaty-like status that has made it increasingly difficult to establish a viable system of marine management without Indonesian agreement. Indonesian authorities have in the past acquiesced in supporting Australian efforts when presented with clear evidence of serious depletion. This was eventually the case with the closure of Ashmore and Cartier to trepang and trochus gathering.

To present reliable evidence on resource depletion, however, the Australian government must monitor resources and establish data on depletion over a period of years. In the case of Scott Reef, where the threat of depletion is greatest, this has only just begun. The first credible study of its kind was carried out only in September 2008.

Fifth, the establishment of the MOU Box in 1989, as opposed to 12-nautical-mile limits around particular reefs, opened Australia to an increase in illegal fishing, because the MOU Box could be used as a strategic departure point for incursions to the south and west. These incursions are, however, only part of a larger problem of illegal fishing from many points of departure along Australia's border. It is this problem that Navy and Customs have had to address over the past several years and must continue to monitor in the future.

Sixth, as part of its 1989 negotiations with Indonesia, the Australian government agreed 'to make arrangements for cooperation in developing alternative income projects in Eastern Indonesia for traditional fishermen traditionally engaged in fishing under the MOU'. Since this time, there have been only two pilot projects to develop alternative livelihoods on Rote and the Bay of Kupang.[24] If such efforts are to succeed, they will require long-term development, especially of those components that have been successfully developed in previous pilot projects.

Finally, it needs to be pointed out that a major source of dispute over Indonesian fishing does not concern the MOU Box at all, but rather the area of sea between two defined boundaries: Australia's seabed boundary and its fishing zone boundary. Roughly a dozen motorized trepang boats were apprehended in 2008 in the zone between these boundaries —

---

24  The Research School of Pacific and Asian Studies at the ANU was involved in one of these projects. It promoted the development of seaweed cultivation to establish an alternative income steam for fishing communities on Rote and experimented with sponge cultivation trials in the Bay of Kupang as another possible livelihood enterprise.

though the fishermen themselves insist that they were in Indonesian waters. Apart from any argument over specific location, the disputed issue is whether the boats were trap fishing, which is permitted, or taking trepang from the seabed, which is prohibited. The captains of four of these vessels were flown from Kupang to Darwin in August 2008 for a hearing of their cases. A decision of the Darwin magistrate may be the next critical judgment to feed into the continuing international dialogue on a longstanding relationship.

## REFERENCES

DFAT (Department of Foreign Affairs and Trade) (n.d.), 'The control of Indonesian traditional fishing in the Australian fishing zone off north-west Australia', unpublished collection of documents, Canberra.

Environment Australia (2002), *Ashmore Reef National Nature Reserve and Cartier Island Marine Reserve (Commonwealth Waters) Management Plans*, Canberra, available at http://www.environment.gov.au/coasts/mpa/publications/pubs/cartier-plan.pdf.

Fegan, Brian (1999), 'Field report on September 1999 socioeconomic research in Probolinggo, East Java', CSIRO project on biology, fishery assessment and management of shared snapper fisheries in northern Australia and eastern Indonesia, Project No. FIS/97 165, Canberra.

Fox, James J. (1977), 'Notes on the southern voyages and settlements of the Sama–Bajau', *Bijdragen tot de Taal-, Land- en Volkenkunde*, 133: 459–65.

Fox, James J. (1992), 'Report on eastern Indonesian fishermen in Darwin', *Illegal Entry*, Occasional Paper Series No. 1, Centre for Southeast Asian Studies, Northern Territory University, Darwin, pp. 13–24.

Fox, James J. (1998), 'Shoals and reefs in Australia–Indonesia relations: traditional Indonesian fishermen', in A. Milner and M. Quilty (eds), *Australia in Asia: Episodes*, Oxford University Press, Melbourne, pp. 111–39.

Fox, James J. (2000), 'Maritime communities in the Timor and Arafura region: some historical and anthropological perspectives', in S. O'Connor and P. Veth (eds), *East of Wallace's Line: Studies of the Past and Present Maritime Cultures of the Indo-Pacific Region*, A.A. Balkema, Rotterdam, pp. 337–56.

Fox, James J. (2008), 'From one coast to the other: episodes in the history of relations between the coasts of Indonesia and Australia', plenary paper presented to the Coast to Coast Conference, Darwin, 18–22 August.

Fox, James J. and Mark Meekan (2006), 'The shark species targeted and caught by Indonesian fishermen in Australian waters', report for the Australian Fisheries Management Agency, Canberra.

Fox, James J. and Sevaly Sen (2002), 'A study of socio-economic issues facing traditional Indonesian fishers who access the MOU Box', report for Environment Australia, Canberra, October, available at http://rspas.anu.edu.au/people/personal/foxxj_rspas/Fishermen_MOU_BOX.pdf.

Fox, James J., Dedi S. Adhuri, Tom Therik and Michelle Carnegie (2009), 'Searching for a livelihood: the dilemma of small-boat fishermen in eastern Indonesia' in B. Resosudarmo and F. Jotzo (eds), *Working with Nature against Poverty: Development, Resources and the Environment in Eastern Indonesia*, Institute of Southeast Asian Studies, Singapore, pp. 201–25.

Horridge, Adrian (1979), *The Lambo or Prahu Bot: A Western Ship in an Eastern Setting*, Maritime Monographs and Reports 39, Trustees of the National Maritime Museum, Greenwich.

Macknight, C.C. (1976), *The Voyage to Marege: Macassan Trepangers in Northern Australia*, Melbourne University Press, Melbourne.

Mellefont, Jeffrey (1988), 'Australian National Maritime Museum', *Indian Ocean Review*, 1(4): 3, 21–4.

Mellefont, Jeffrey (1991), 'Between tradition and change: an Indonesian *perahu* in an Australian collection', *Great Circle*, 13(2): 97–110.

Resosudarmo, Budy, Lydia Napitupulu and David Campbell (2009), 'Illegal fishing in the Arafura Sea', in B. Resosudarmo and F. Jotzo (eds), *Working with Nature against Poverty: Development, Resources and the Environment in Eastern Indonesia*, Institute of Southeast Asian Studies, Singapore, pp. 178–200.

Russell, Barry C. and Lyle L. Vail (1988), 'Report on traditional Indonesian fishing activities at Ashmore Reef Nature Reserve', report for the Australian National Parks and Wildlife Service, Canberra.

Stacey, Natasha (1992), 'The *Tujuan*: the study of the material culture of an Indonesian fishing vessel held in the collection of the Northern Territory Museum of Arts and Sciences', Graduate Diploma thesis, James Cook University, Townsville.

Stacey, Natasha (2007), *Boats to Burn: Bajo Fishing Activity in the Australian Fishing Zone*, Asia-Pacific Environment Monograph 2, ANU E Press, Canberra, available at http://epress.anu.edu.au.

Wallace, Alfred Russel (1869), *The Malay Archipelago*, Volume 2, Macmillan, London, reprinted in 1986 by Oxford University Press.

# 13 FLUID BOUNDARIES: MODERNITY, NATION AND IDENTITY IN THE RIAU ISLANDS

*Michele Ford and Lenore Lyons*

The Indonesian language equivalent for the word 'fatherland' ... is 'tanah air' meaning 'land-water', thereby indicating how inseparable the relationship is between water and land to the Indonesian people. The seas, to our mind, do not separate but connect islands. More than that, these waters unify our nation (Indonesian delegation to UNCLOS III, cited in Puspitawati 2005: 2–3).

The archipelagic concept (*wawasan nusantara*) has been central to Indonesian nation building, because the concept of Indonesia is predicated on clear territorial boundaries that encompass both land (*tanah*) and water (*air*). This concept was first articulated through the Juanda Declaration of 1957 (see Chapter 2 by Butcher) and received further legitimacy when Indonesia's status as an archipelagic state was recognized under the United Nations Convention on the Law of the Sea (UNCLOS). Whereas the concept of archipelagic statehood is concerned with ensuring national territorial integrity (that is, it is outwardly oriented), *wawasan nusantara* is focused on the internal dynamics of national integration in an archipelagic nation characterized by ethno-linguistic diversity. The archipelagic concept and the archipelagic state are nevertheless intrinsically connected, as each relies on the other for its legitimacy. Fundamental to both is the view that the sea unites Indonesia's islands and the people living on them. This idea is expressed in comments such as *laut adalah perekat kepulauan Indonesia* (the sea is the glue of the Indonesian archipelago) (Adhuri 2003: 4). In other words, the seas located within the territorial baselines that surround the archipelago draw the people of Indonesia together to form one, united nation, just as the international maritime border which marks out the edges of the archipelagic state serves to separate the Indonesian nation and its people from other nations. The archipelagic concept emerged in the immediate post-independence period as Indonesia's leaders faced the challenge of encouraging Indonesians to think of themselves as a nation. To achieve this goal, the idea of the Indonesian nation was intensified and redeployed through a range of state ideologies, including *wawasan nusantara*. National development projects have been especially important in realizing the archipelagic concept (cf. Barker 2005). The concept has also been reinforced through official statements and rituals, the media, the education system and the practices of the bureaucracy and the military. But just as it took many years for the concept of the archipelagic state to achieve international recognition and acceptance, so too has it taken time for the idea of a single *tanah air*, or land united by water, to become accepted in the way that Indonesians experience and imagine the nation. *Wawasan nusantara*, like other forms of national imagining, remains a process in the making rather than a statement of fact, and Indonesia's territorial seas, rather than being a source of unity, continue to divide many Indonesians (Adhuri 2003).

The archipelagic concept continues to be challenged not only by internal divisions based on ethnicity and culture, but also by the arbitrariness of the international maritime border that demarcates the edges of the archipelagic state. The Riau Islands, which are scattered across the Malacca Strait to the northeast of Sumatra and directly south of Singapore, are a case in point (see Map 13.1).[1] The maritime borders that separate the main islands of Bintan, Batam and Karimun from Singapore and Malaysia delineate part of the northern boundary of the archipelagic state. However, for the many islanders who feel a strong sense of connection to Singapore and Malaysia, these borders have limited practical or symbolic significance. The proximity of these places means that the seas, rather than uniting Riau Islanders with other Indonesians in far-flung parts of the archipelago (as implied by the archipelagic concept), act as

---

1  The province of Kepulauan Riau (Kepri) consists of the islands of Bintan, Batam and Karimun, those of the Natuna and Lingga archipelagos and many other smaller islands and islets. We use the term 'Riau Islands' here to refer to the three main islands of Bintan, Batam and Karimun that lie to the south of the Malacca Strait. The research on which this paper is based was funded by an Australian Research Council (ARC) Discovery Project grant for the project, 'In the shadow of Singapore: the limits of transnationalism in insular Riau' (DP0557368). The component of the fieldwork that focused on the Riau Islands included ethnographic observation and over a hundred semi-structured interviews with Riau Islanders of different backgrounds. Most of the interviews were conducted in the port cities of Tanjung Pinang and Tanjung Balai Karimun.

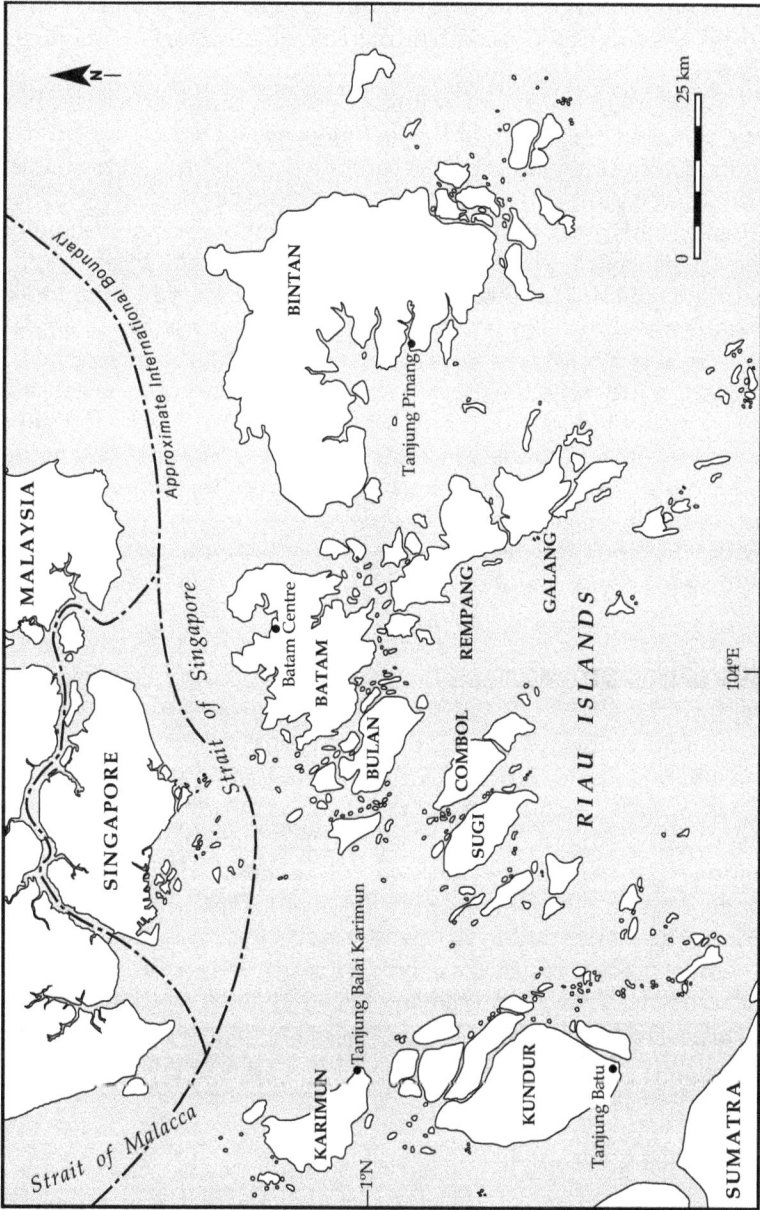

Map 13.1    *The Riau Islands*

a conduit between them and their neighbours. It also creates a sense of regional identity at odds with the concept of *wawasan nusantara* that hints at the multiple ways in which the seas, the archipelago and the nation are imagined and experienced by different communities within the nation-state.

While there is growing scholarly interest in the everyday meanings and practices of nation building in Indonesia, there has been little consideration of the role the sea plays in drawing the nation together or in the ways that it pulls it apart.[2] Even less attention has been given to the ways in which the communities that live along Indonesia's maritime borders experience and understand the border in their daily lives.[3] By focusing on the communities that live along the outer edges of the Indonesian nation-state, this chapter offers a means to explore the everyday salience of water for the process of constructing the *wawasan nusantara*. We argue that although Indonesia's maritime borders are a constant signifier of national identity and belonging for Riau Islanders, at the same time islanders have a strong sense of regional identity and belonging built on a dense web of transnational flows across the straits. Our research reveals that, for many Riau Islanders, it is the border that unites them with other Indonesians, but it is the waters of the straits that draw them to their northern neighbours.

## DIVIDING THE WATERS

Like many borders, the territorial line that divides Indonesia from Malaysia and Singapore cannot be seen. The contemporary maritime boundaries that mark territorial spaces in the Malacca Strait are based on different legal systems and rights of passage arising from moves by Malaysia and Indonesia to extend their maritime boundaries during the 1970s. The Singapore Strait, which separates Singapore from the Riau Islands, is just 3.2 nautical miles at its narrowest and is constantly less than 15 nautical miles across. As Malaysia and Indonesia each claim 12 nautical miles of territorial waters, and Singapore claims three nautical miles, neither strait is sufficiently wide to accommodate their territorial claims. As a result, treaties have been negotiated to locate the international border at the mid-point (Roach 2005). Although the territorial border between

2   One exception is the working paper by Adhuri (2003). For a different way of conceiving of the archipelago, see Boellstorff's (2005) work, *The Gay Archipelago*, which considers the intersections between sexuality, nation and identity.
3   The exception is the seafarers who fish in the waters between Indonesia and Australia; see Chapter 12 by Fox.

Indonesia and Singapore is considered to be a fairly stable marker of territorial sovereignty, some claims remain unresolved (Mak 2006). Agreement was recently reached on the western boundary of the border but negotiations continue on the eastern boundary (Osman 2009).

The maritime borders that divide Indonesia from its northern neighbours are based on a line originally drawn up between the British and Dutch colonial powers in 1824 as part of the Anglo-Dutch Treaty, which gave the British the Malay Peninsula and the island of Singapore at its tip and the Dutch Sumatra, including the Riau Islands. This division did not draw on pre-existing cultural or political boundaries. Rather, it reflected desired colonial spheres of trade and influence. Although it effectively split the Johor–Riau Sultanate in half, the original treaty was not a right to possession or governance, but simply a right to influence.[4] With time, however, the 'line of demarcation' between the two colonial powers came to represent the border between their respective colonial territories of British Malaya and Singapore and the Netherlands East Indies and, much later, between three separate countries. The border as it stands today took shape over some decades. Indonesia declared independence in 1945, but the Riau Islands were not incorporated administratively into the republic until the Dutch finally acceded to Indonesia's claim to independence in 1949. Across the straits, the Singapore and Malaysian borders were not established until Singapore declared independence in 1965 after a short-lived merger in the Federation of Malaysia in 1963.

These international borders were largely ineffective in dividing the waters between the Riau Islands and Singapore and Malaysia before the late twentieth century. Throughout the colonial period and up to the 1960s, individuals crossed the straits regularly with little regard for the markers of territorial sovereignty or jurisdiction. In doing so, they were following well-travelled trade routes established during pre-colonial times by Malay and Bugis traders, and further strengthened by the presence of Chinese migrants, who began moving into the region in the 1800s. As Singapore transformed itself into the dominant politico-economic power in the region, the Riau Islands became increasingly tied to its economy (Trocki 1990) and trading networks flourished across the straits. It was only with Confrontation, when Indonesia's northern periphery was a key site of skirmishes between Indonesia and Malaysia, that the concepts of citizenship and nationality became more closely linked to notions of sovereignty and boundary maintenance. Thus Confrontation, rather than Indonesian independence, was pivotal in reimagining and reconstructing the borders in the straits (Ford and Lyons 2006).

---

4  For a discussion of the relationship between the Johor–Riau Sultanate and the colonial powers in the region, see Trocki (1979).

By enclosing sovereign territory, the newly enforced border not only restricted movement across the straits, but also made people on either side of the border into national citizens whose rights of passage were determined by the possession of a passport. As immigration and customs regimes became more entrenched in the decades that followed, the relative freedom of mobility that had characterized the crossing of the straits during the early years of Indonesian independence began to change. Since this time, Riau Islanders have continued to cross the straits but with a heightened awareness of moving from one country to another.

## WATERS THAT UNITE AND DIVIDE

The composition of the Riau Islands' population reflects the constant migratory flows of different populations into and out of the region over the last century. The original population consisted of Orang Melayu (Malays) and Orang Laut (nomadic fisher-people) (Wee 1985). During the colonial period, considerable numbers of Bugis traders and warriors, as well as Chinese traders and coolies, began to arrive, creating an increasingly diverse and vibrant community (Roeroe et al. 2003). In 1930, 60 per cent of Tanjung Pinang's population was Chinese (Butar-Butar 2000: 5). However, the influx of migrants from Java and Sumatra over the last three decades has shifted the ethnic balance between the Chinese and non-Chinese in the city – and to a lesser extent in other parts of the Riau Islands – and created an even more ethnically diverse population. Figures from the 2000 census show that on Batam, the Javanese constitute the single largest ethnic group, followed by Malays, Minang and Bataks. On Bintan, Malays make up the largest ethnic group, followed by Javanese and Minang. The Chinese community comprises just under 10 per cent of the population of the entire province of Kepulauan Riau (Ananta and Bakhtiar 2005: 20).

Local Malays, the islands' large Chinese population and many longterm residents from other ethnic groups have close and enduring links across the straits. Many Chinese and Malay families living in the islands have relatives in Singapore and Malaysia. Wealthier residents seek healthcare in Malacca and Kuala Lumpur and educate their children abroad, and since regional autonomy, Singapore government departments, educational institutions and cultural groups have initiated a wide range of cross-border outreach programs. Middle-class Riau Islanders also travel across the straits for leisure, particularly shopping and entertainment. Many other Indonesians cross the border to work either on a long-term basis on plantations or in secondary industry or services in Malaysia, or for shorter stints in Singapore. Significant numbers of internal migrants who move to Batam, Bintan or Karimun in search of work in the industrial zones end up working overseas, and most international migrants spend long periods of time in the islands prior to departure, or return there between placements (Lyons and Ford 2007). Some skilled workers employed in Singapore-owned factories in the industrial enclaves of Batam and Bintan spend time working in the Singapore premises of parent companies, while other workers move across the border on a daily or weekly basis, mostly illegally. These short-term labour migrants work as traders in second-hand whitegoods and computers, handymen or seamstresses, but also in restaurants and as sex workers.

Singapore's considerable investment in the islands under the Indonesia–Malaysia–Singapore (IMS) Growth Triangle initiative means that Singapore businesspeople and company employees also move back and forward across the straits. Meanwhile, travel agents in Singapore promote the luxury resorts on Bintan and Batam as local, rather than international, holiday destinations (Ford and Lyons 2006). In addition, thousands of working-class Singaporean and Malaysian men travel to the islands each month for sex (Ford and Lyons 2008). Some of these men establish first or second households in the islands, joining other groups involved in cross-border relationships or marriages, such as those with longstanding cross-border family ties and individuals with experience working in Singapore and Malaysia. The physical proximity of Singapore, peninsular Malaysia and the Riau Islands enables couples who live on opposite sides of the border to see each other regularly while maintaining separate lives in their home countries (Lyons and Ford 2008).

This is not to suggest that all Riau Islanders—or all Singaporeans, for that matter—have the same experience of the border. An individual's exposure to transnational encounters is influenced by his or her proximity to an international port. Only a limited number of official international ports service the cross-border ferries on each island. Whereas unauthorized ports (*pelabuhan tikus*, literally 'rat ports') also facilitate cross-border movement, the majority of travellers cross the border through the official ports. The important role of these ports as exit and entry points means that different communities within a short distance of each other may have dramatically different experiences of the border, even if both are located close to the sea. Islanders living in the port towns of Tanjung Balai Karimun and Tanjung Pinang or on the island of Batam have ready access to international ferry terminals. In contrast, islanders living near Bintan's deep-water port of Kijang, which services Indonesian cargo ships and domestic Pelni routes, have more limited opportunities to cross the border. Residents of Kijang and villages on the east and north of the island must first travel to Tanjung Pinang (usually overland) if they wish to travel to Singapore or Malaysia. For this reason, they are less likely to have crossed the straits. Conversely, for people living in Tanjung Balai Karimun, ferries to Singapore and Malaysia are both cheaper and faster than the interisland ferry to Bintan. As long as they hold a passport and make a day trip or can stay with friends or relatives, both these countries are more accessible physically, and even financially, to residents of Karimun than is Tanjung Pinang, the provincial capital.[5]

Yet even with easy access to an international port, residents of port towns or those living elsewhere in the islands who have the desire and means to travel may not succeed in entering the territorial space of Singapore or Malaysia. During the 1990s, the IMS Growth Triangle was often cited as an example of an increasingly 'borderless' world in which people, goods and information flowed seamlessly across national borders (Ohmae 1990, 1995). But as critics of this view argue, nation-states continue to play powerful roles in territorializing global order, and individuals who cross international borders continue to be constructed as national subjects (van Houtum and van Naerssen 2002; Cunningham 2004). Thus, although Indonesians, Malaysians and Singaporeans officially have reciprocal rights to visa-free tourism, these rights are treated differently at various border control sites. Similarly, although all three countries require work visas, this requirement is policed in different ways.

As Heyman (2004) notes in his work on international airports, immigration checkpoints, which are usually located within the territorial boundaries of the nation-state, have become the border gates of the nation. While Singaporeans and Malaysians pass through Indonesian checkpoints regularly without being questioned, every Indonesian who seeks to enter Singapore or Malaysia faces a greater or lesser risk of being turned back. Sometimes refusal of entry is imposed arbitrarily. The treatment accorded Riau Islanders at immigration checkpoints is also affected by ethnicity, age and class. Wealthy people are less likely to be refused visas or asked to show that they have sufficient funds to cover the cost of travel abroad. Social status and money do not always guarantee an easy passage, however. Our research suggests that Chinese Indonesians of all classes are able to enter Singapore more easily than Indonesians from other ethnic backgrounds, and that older people of all backgrounds have less trouble crossing either border.

The disparities between the ways Indonesians and their cross-border counterparts experience immigration regimes are nowhere more visible than on the Singapore–Indonesia border, where Riau Islanders have great difficulty crossing into Singapore because of Singapore government con-

---

5   Residents of Bintan, Batam and Karimun are exempt from the exit tax levied on Indonesians travelling internationally.

cerns that they may engage in illegal work or overstay their tourist visas.[6] As unemployment grew in the islands in the wake of the Asian financial crisis, both Singapore and Malaysia stepped up surveillance to prevent unauthorized labour migrants from attempting to cross the border with or without passports in search of work.[7] Increased pressure on the border led to heightened immigration controls at the major checkpoints in the Harbour Front and Tanah Merah terminals in Singapore used by ferry services from the Riau Islands. These efforts appear to have had some success in reducing the numbers of people attempting to enter Singapore by sea (Ministry of Home Affairs 2007a, 2007b). Malaysia's attempts to stem the flow of undocumented labour migrants from Indonesia have been much less successful, in part because of a lack of commitment and wherewithal on the Malaysian side. Hundreds of thousands of Indonesians cross into Malaysia without proper documentation, but these irregular migration flows are typically dealt with through forced repatriation of undocumented labour migrants rather than by interception at the border (Ford 2006).

The proximity of Singapore and Malaysia means that the islands are also a strategic hub for people-smuggling syndicates, which operate throughout Indonesia (Agustinanto 2003: 178). In recent years, the provincial government has passed human trafficking regulations and local police departments have established countertrafficking desks. International agencies such as the International Organization for Migration (IOM) have provided support for these efforts, leading to increasing numbers of arrests of 'traffickers' and the identification of significant numbers of 'victims' — the majority of whom are migrant workers (IOM 2007). Counterintuitively, it is therefore generally easier for young men to cross the border than young women, because immigration and customs officials in Singapore (and to some extent Malaysia) are conscious of the concerns surrounding the trafficking of young women.

The security responses to these illegal border crossings have converged with the increasingly visible presence of navy and customs boats as a result of bilateral and multilateral agreements and joint policing initiatives on piracy between Indonesia, Malaysia and Singapore.[8] The Malacca Strait is one of the world's busiest waterways. More than 50,000

---

6  For an extended discussion of differential access across the Singapore–Indonesia border and how it has changed over the decades, see Ford and Lyons (2006).

7  While some of these 'illegal migrants' travelled to Singapore and Malaysia from other parts of Indonesia, many were initially drawn to Batam through the promise of work in the industrial zones established under the IMS Growth Triangle initiative (Lyons and Ford 2007).

8  For a discussion of these initiatives, see Ong-Webb (2006).

ships pass through it each year, carrying more than one-third of global trade and two-thirds of the world's liquefied natural gas (Ong-Webb 2006: xix). The strait has strategic importance for Singapore because its economy is largely dependent on the shipping trade: if the Singapore Strait were closed, ships would have to travel an additional 500 nautical miles, with an immediate effect on freight costs worldwide (Roach 2005). By contrast, several commentators have pointed out that Indonesia is much less invested in securing the strait, because it accounts for only a small proportion of the country's sea trade (Teo 2007).

Piracy, which thrived during the pre-colonial and colonial eras (Manap 1983; Tagliacozzo 2007), has long been a matter of international concern.[9] The topography of the region, with its numerous small islands covered in dense mangroves and dissected by tidal streams, offers protection to pirates who are able to take advantage of their knowledge of the land and sea to escape detection. In 2005, one-quarter of the world's reported piracy attacks occurred in Southeast Asia — a majority of them in the Malacca Strait and in the territorial waters of Indonesia (Ong-Webb 2006).[10] Scholars who have studied piracy in the region, such as Frécon (2006), have found few links between piracy and terrorism, arguing that in the Riau Islands piracy is, rather, an opportunistic activity linked to petty crime. However, the international view that piracy and terrorism are linked appears to have achieved traction at the regional level in the post-9/11 world. In June 2005, the Joint War Committee of Lloyd's Market Association described the Malacca Strait as an area at risk of 'war, strikes, terrorism and related perils' due to the 'intensification of the weaponry and techniques used by the pirates in the Straits … who are now largely indistinguishable from terrorists' (cited in Teo 2007: 541).

The narrowness of the straits, combined with their role as a key international shipping lane, has given rise to concerns about terrorist activities, including 'floating bombs' that could be used to target Singapore's harbour and oil refineries (Vijayan 2004). Fears of terrorism also extend

---

9    The world's international maritime powers have proposed a range of initiatives to address security in the straits. However, the littoral states are concerned that international initiatives to secure safe passage may undermine their sovereign rights. While UNCLOS defines piracy as an act that occurs on the high seas or outside the jurisdiction of any state, the International Maritime Bureau has changed its definition to allow acts of violence against ships in national waters to be treated as piracy on the high seas. This means that one state may take action against pirates in the national waters of another state, with consequences for national sovereignty.

10   There is evidence that the number of reported attacks on international ships has declined since 2005, although attacks on fishing vessels (often not reported) continue (Murphy 2007).

to rumours of possible bombings in the Singaporean tourist zone in the north of Bintan, exemplified by reports that Bintan could become 'another Bali ... only closer to home'.[11] These concerns shape the way the straits (and the islands) are perceived regionally and internationally. The international perception that the straits are a dangerous place inhabited by pirates, traffickers and terrorists, and the security forces that seek to control them, has had a very real impact on the ability of Riau Islanders to cross the border and thus has affected the sense of connection that Riau Islanders feel with their near neighbours.

## IDENTITY AND NATION IN THE BORDER ZONE

Despite the increasing securitization of the border, Singapore and Malaysia remain central in local imaginings of identity, nationhood and belonging. For Riau Islanders living in the port towns of Tanjung Pinang and Tanjung Balai Karimun and anywhere on Batam, the trappings and rituals of the international ports and the foreigners who enter their communities through them are part of their daily geography. Individuals do not need to cross the border to feel a sense of connection with Malaysia and Singapore, because they regularly engage in transnational encounters with commuters and other foreigners who purchase supplies in their shops, eat in their food stalls and walk on their streets. Whether or not Riau Islanders wish to, can afford to or are able to travel across the border, the proximity of Singapore and Malaysia, and the visibility of Singaporeans and Malaysians in their towns, on television and in their newspapers, makes those places seem easy to know.

This is not to suggest that those who are less mobile have a more uniform experience of transnationalism than those who cross the border. The wealthy businessman who takes his family to Malaysia for regular holidays and the Chinese market stall owner who saves every penny to educate her children in Singapore experience transnationalism differently from the second-hand goods trader who crosses the border each month to buy new stock. Similarly, the sex worker employed in the busy port town brothel, the motorcycle taxi driver who hustles tourists on the dock and the immigration official who stamps passports in the arrival hall have a more intense transnational experience than the primary school teacher who has neither the means nor the desire to cross the border or to engage with foreigners. Ultimately, however, it is impossible to live in any but the most isolated communities on Batam, Bintan and

---

11 'Shock and disbelief at attack so close to home', *Straits Times*, 15 October 2002.

Karimun without developing some awareness of the 'other' across Indonesia's northern maritime borders, even if simply to reject the Singapore rat race in favour of the more 'civilized' pace of life at home. The cross-straits connections that Riau Islanders have forged and the constant flows of people and goods across the border have thus created a way of life that is different to that which they imagine other Indonesians experience. This sense of the special character of islander life emerges in a context where travel to other parts of Indonesia is relatively rare for long-term residents. The tyranny of distance has lessened considerably with the deregulation of the airline industry and subsequent advent of cheaper airfares and a wider array of destinations. A limited air service has been re-established between Tanjung Pinang and Pekanbaru, the capital of mainland Riau, but Batam remains the only place in the islands effectively linked to the rest of Indonesia by air. The cost of tickets and the need to travel to Hang Nadim airport in Batam means that air travel is still beyond the reach of most islanders. Water therefore continues to provide an important conduit between the islands and Pekanbaru, and between the islands and Jakarta. However, the sea journey to these other parts of Indonesia is time consuming and arduous. The journey to Jakarta takes 24 hours on the Pelni boats that dock in the port of Kijang on Bintan. Other large and medium-sized ships link Batam and Karimun with the major north Sumatran port of Belawan, outside Medan, and Padang in West Sumatra. A journey by boat to Pekanbaru entails either a journey up the Siak River in a small motorized vessel, which takes three days and two nights, or several hours by boat to the mainland port of Dumai, followed by several more overland. By contrast, the journey from Batam or the northern reaches of Bintan to Singapore is less than an hour by ferry. Karimun, which lies to the west, is also less than an hour by ferry from the Malaysian state of Johor, and little further to Singapore.

Islanders' sense of the physical distance between the Riau Islands and other parts of Indonesia has diminished somewhat in the last two decades with the influx of migrant workers attracted to the islands by the prospect of work in the IMS Growth Triangle. These vast new flows of migrants have irrevocably changed the character of island communities, since the newcomers have not assimilated into the local community to the extent that their predecessors did (Ford 2003). Improvements in information and communication technology have also helped to integrate the Riau Islands into the Indonesian nation and make island communities less outward looking (Faucher 2006), and with the exception of the longstanding Chinese community—many of whose members speak little Indonesian—most islanders now watch Indonesian television broadcasts, whereas just 20 years ago Singaporean and Malaysian stations were the only ones Riau Islanders could watch. However, this growing awareness of their Indonesianness has by no means displaced Riau Islanders' sense of connection across the straits. In places like Tanjung Pinang and Tanjung Balai Karimun, where there are strong, traditional community links across the border, islanders emphasize the Riau Islands' 'shared history' and 'family links' with Singapore and Malaysia. In describing their sense of belonging and identity, local Malays continue to refer to the longstanding kinship and trade ties that bound the Riau Islands and Singapore—and to a lesser extent Johor—during the colonial period and since (Ford and Lyons 2006). They argue that these networks are the basis for a regional identity shared by those who live along the border. These accounts resonate with the discourse of 'pan-Malayness' that is sometimes invoked in scholarly and popular accounts of the region (cf. Barnard 2004) and is common in the scholarly literature on the Riau Islands (Wee 1985, 2002; Wee and Chou 1997; Benjamin and Chou 2002).

Writing about the salience of such a Pan-Malay identity in contemporary Singapore and Malaysia, Kahn (2006: 82) claims that 'it is in the modern trans-border Malay World that one is most likely to find genuine alternatives to nationalist narratives, as well as the sources of resistance to the projects of building modern states and nations in the region'. Similarly, Rahim (1998: 16) argues that pan-Malay consciousness continues to have potency, as demonstrated by the cultural, social and economic links that Singaporean Malays are forging at a regional level. However, as Cribb and Narangoa (2004) suggest—citing the example of the 'Malays' in Malaysia and Indonesia—transnational identities are difficult to maintain in the face of political boundaries and international borders that are strengthened through a range of state practices. This is borne out by our research among young Malay Singaporean men, who say that they feel little connection with 'other Malays', and even less with Indonesians (Lyons and Ford, forthcoming). And, with the exception of some among the Malay elites, almost none of the Riau Islanders we spoke to explicitly invoked any kind of overarching pan-Malay identity in describing their connections with communities on the other side of the maritime border. Instead, they argued that the waters of the straits drew them together and shaped a common sense of place and destiny. This suggests that contemporary forms of cross-border identity are the product of location rather than ethnicity or culture, emerging out of the increasing frequency and volume of transnational flows and individual movements across the straits (and beyond).

This sense of regional belonging described by many Riau Islanders has been enhanced by the efforts of the central government to incorporate the islands into Indonesia's national project. The IMS Growth Triangle and the location of the Riau Islands so close to Indonesia's neighbours means that the government has increasingly sought to regulate the region through a strong bureaucratic and navy presence, often overriding the insights and knowledge of local government and local communities. Riau Islanders are acutely aware, and often resentful, of central government initiatives that impinge on their way of life, leading them to complain constantly of central bureaucrats' lack of understanding of local conditions — that the central government and its representatives simply 'do not understand' conditions in the islands. They have been equally wary of attempts to draw them into the political concerns of those living on mainland Sumatra. When Malay elites in Pekanbaru attempted to establish a separatist movement soon after the fall of Soeharto, they did so with almost no support from the Riau Islands. The successful push in the islands for the establishment of an independent province in the early 2000s further emphasized their sense of difference from mainlanders, regardless of ethnic background (Ford 2003).

The islanders' close cross-border connections do not mean, however, that they feel a stronger sense of connection with the governments in Singapore or Kuala Lumpur than they do with Jakarta. Nor do Riau Islanders imagine themselves as part of a post-national community. Their sense of their place in the world is firmly grounded in their view of themselves as both Riau Islanders and Indonesians. They see their way of life as being intimately tied to the lives of their neighbours, but their sense of regional belonging also stands in opposition to their view of Singaporeans and Malaysians as different to themselves. The dramatic devaluation of the rupiah following the Asian financial crisis of 1997–98 and the growing differences in the purchasing power and lifestyles of islanders compared with Singaporeans and Malaysians have further served to emphasize the dividing effects of the border. The experiences of urban Riau Islanders living in and around the main port towns on Bintan, Batam and Karimun thus do not make the existence of the Indonesian state redundant or irrelevant. Their sense of identity and belonging is defined by their location on the northern edges of the Indonesian nation-state. This awareness is built on a level of understanding that precludes blind admiration or acceptance of the values of their 'more developed' neighbours. At the same time as a starry-eyed factory worker stares at the glittering Singapore skyline clearly visible from Batam's north coast, a sex worker in Tanjung Pinang speaks scornfully of the background and lifestyle of a working-class Malaysian client, and a Chinese businessman in Tanjung Balai Karimun reflects pityingly on the rat race his Singaporean cousin must endure.

Indonesia's maritime borders are therefore a constant presence in the consciousness of those who live in the islands. These invisible borders, made present in the border practices of authorities in Malaysia and Singapore and reinscribed through the actions of the navy and customs officials from all three countries who patrol the straits, mark out their identity as 'Indonesians'. Their knowledge of their Indonesianness is made evident in the need to possess a passport to cross the waters that divide them from their neighbours and is reinforced through the actions of a central government that seeks to include Riau Islanders in its nation-building agenda through the provision of national education and telecommunications systems. It is also ever-present in the behaviour and actions of recent migrants, whose sense of place is oriented away from the border, and in the lifestyles of Singaporeans and Malaysians, the majority of whom share little in common with their islander neighbours.

At the same time, Riau Islanders have a strong sense of their place in an archipelago — not the Indonesian archipelago invoked by the concept of *wawasan nusantara*, but an imagined archipelago of islands drawn together by the waters of the Malacca Strait. This imagined space cannot be mapped neatly against any known archipelagic zone (such as the Riau Archipelago) or the existing provincial boundary (Kepulauan Riau). It is an imagined space that transcends the international border and signifies the dense web of transnational linkages between the differently situated communities living along Indonesia's maritime edge. In professing a sense of regional belonging and identity, Riau Islanders do not pretend that this sense of place is experienced by all who live along the border. However, it is central to the way in which they think about their place within the Indonesian nation, for while the international border ties Riau Islanders to Indonesia, the sea ties them to the lands across the straits.

## REFERENCES

Adhuri, Dedi Supriadi (2003), *Does the Sea Divide or Unite Indonesians? Ethnicity and Regionalism from a Maritime Perspective*, Australian National University, Canberra.

Agustinanto, Fatimana (2003), 'Riau', in R. Rosenberg (ed.), *Trafficking of Women and Children in Indonesia*, International Catholic Migration Commission and American Center for International Labor Solidarity, Jakarta, pp. 178–82.

Ananta, Aris and Bakhtiar (2005), 'Who are "the lower class" in Riau Archipelago, Indonesia?', paper presented to the Sixth Annual Population Researcher Conference on Linkage between Population and Millennium Development Goals: The Asian Perspective, 29 November – 1 December, Islamabad.

Barker, Joshua (2005), 'Engineers and political dreams: Indonesia in the satellite age', *Current Anthropology*, 46(5): 703–27.

Barnard, Timothy P. (ed.) (2004), *Contesting Malayness: Malay Identity across Boundaries*, Singapore University Press, Singapore.

Benjamin, Geoffrey and Cynthia Chou (2002), *Tribal Communities in the Malay World: Historical, Cultural and Social Perspectives*, International Institute for

Asian Studies in conjunction with the Institute of Southeast Asian Studies, Leiden and Singapore.

Boellstorff, Tom (2005), *The Gay Archipelago: Sexuality and Nation in Indonesia*, Princeton University Press, Princeton NJ.

Butar-Butar, V. (2000), 'Dinamika sosial politik di propinsi Riau: catatan seorang awam' [Socio-political dynamics in Riau province: notes of an ordinary person], paper presented to an international seminar on Dinamika Politik Lokal di Indonesia: Perubahan, Tantangan dan Harapan [Local political dynamics in Indonesia: changes, challenges and expectations], Yogyakarta, 3–7 July.

Cribb, Robert and Li Narangoa (2004), 'Orphans of empire: divided peoples, dilemmas of identity, and old imperial borders in East and Southeast Asia', *Comparative Studies in Society and History*, 46(1): 164–87.

Cunningham, Hilary (2004), 'Nations rebound? Crossing borders in a gated globe', *Identities: Global Studies in Culture and Power*, 11(3): 329–50.

Faucher, Carole (2006), 'Popular discourse on identity politics and decentralisation in Tanjung Pinang public schools', *Asia Pacific Viewpoint*, 47(2): 273–85.

Ford, Michele (2003), 'Who are the *Orang Riau*? Negotiating identity across geographic and ethnic divides', in E. Aspinall and G. Fealy (eds), *Local Power and Politics in Indonesia: Decentralisation and Democratisation*, Institute of Southeast Asian Studies, Singapore, pp. 132–47.

Ford, Michele (2006), 'After Nunukan: the regulation of Indonesian migration to Malaysia', in A. Kaur and I. Metcalfe (eds), *Divided We Move: Mobility, Labour Migration and Border Controls in Asia*, Palgrave Macmillan, New York, pp. 228–47.

Ford, Michele and Lenore Lyons (2006), 'The borders within: mobility and enclosure in the Riau Islands', *Asia Pacific Viewpoint*, 47(2): 257–71.

Ford, Michele and Lenore Lyons (2008), 'Making the best of what you've got: sex work and class mobility in the Riau Islands', in M. Ford and L. Parker (eds), *Women and Work in Indonesia*, Routledge, London and New York, pp. 173–94.

Frécon, Eric (2006), 'Piracy and armed robbery at sea along the Malacca Straits: initial impressions from fieldwork in the Riau Islands', in G.G. Ong-Webb (ed.), *Piracy, Maritime Terrorism and Securing the Malacca Straits*, Institute of Southeast Asian Studies, Singapore, pp. 68–83.

Heyman, Josiah McC. (2004), 'Ports of entry as nodes in the world system', *Identities: Global Studies in Culture and Power*, 11(3): 303–27.

IOM (International Organization for Migration) (2007), *Victims of Trafficking (VoT) Assisted by IOM Indonesia (March 2005 – April 2007)*, Geneva, available at http://www.iom.or.id/statistics.jsp?lang=eng.

Kahn, Joel S. (2006), *Other Malays: Nationalism and Cosmopolitanism in the Modern Malay World*, Singapore University Press, Singapore.

Lyons, Lenore and Michele Ford (2007), 'Where internal and international migration intersect: mobility and the formation of multi-ethnic communities in the Riau Islands transit zone', *International Journal on Multicultural Societies*, 9(2): 236–63.

Lyons, Lenore and Michele Ford (2008), 'Love, sex and the spaces in-between: Kepri wives and their cross-border husbands', *Citizenship Studies*, 12(1): 55–72.

Lyons, Lenore and Michele Ford (forthcoming), 'Singaporean first: challenging the concept of transnational Malay masculinity', in D. Heng and S.M.K. Aljunied (eds), *Reframing Singapore: Memory, Identity and Trans-regionalism*, Amsterdam University Press, Amsterdam.

Mak, J.N. (2006), 'Unilateralism and regionalism: working together and alone in the Malacca Straits', in G.G. Ong-Webb (ed.), *Piracy, Maritime Terrorism and Securing the Malacca Straits*, Institute of Southeast Asian Studies, Singapore, pp. 134–62.

Manap, Normala (1983), 'Pulau Seking: social history and an ethnography', B.Soc.Sc. (Hons) thesis, Department of Sociology, Faculty of Arts and Social Sciences, National University of Singapore, Singapore.

Ministry of Home Affairs (2007a), 'Enforcement against immigration offenders 2001', Singapore, available at http://www.mha.gov.sg/basic_content. aspx?pageid=82, accessed 30 November 2007.

Ministry of Home Affairs (2007b), 'Enforcement against immigration offenders and publicity efforts 1999', Singapore, available at http://www.mha.gov.sg/ basic_content.aspx?pageid=83, accessed 30 November 2007.

Murphy, Martin N. (2007), 'Chapter one: contemporary piracy', *Adelphi Papers*, 47(388): 11–44.

Ohmae, Kenichi (1990) *The Borderless World: Power and Strategy in the Global Marketplace*, Harper Collins, London.

Ohmae, Kenichi (1995) *The End of the Nation State: The Rise of Regional Economies*, Free Press, New York.

Ong-Webb, Graham Gerard (2006), 'Southeast Asian piracy: research and developments', in G.G. Ong-Webb (ed.), *Piracy, Maritime Terrorism and Securing the Malacca Straits*, Institute of Southeast Asian Studies, Singapore, pp. xi–xxxviii.

Osman, Salim (2009), 'Maritime border talks: S'pore, Indonesia agree on western boundary', *Straits Times*, 4 February.

Puspitawati, Dhiana (2005), 'The east/west archipelagic sea lanes passage through the Indonesian archipelago', *Maritime Studies*, 140: 1–13.

Rahim, Lily Zubaidah (1998), *The Singapore Dilemma: The Political and Educational Marginality of the Malay Community*, Oxford University Press, New York.

Roach, J. Ashley (2005), 'Enhancing maritime security in the straits of Malacca and Singapore', *Journal of International Affairs*, 59(1): 97–116.

Roeroe, F., J. Rawis, J. Woek, F. Lumanauw, M. Umbas and N. Lumanauw (2003), *Batam Komitmen Setengah Hati* [Batam: A Half-hearted Commitment], Aksara Karunia, Jakarta.

Tagliacozzo, Eric (2007), *Secret Trades, Porous Borders: Smuggling and States along a Southeast Asian Frontier, 1865–1915*, National University of Singapore Press, Singapore.

Teo, Yun Yun (2007), 'Target Malacca Straits: maritime terrorism in Southeast Asia', *Studies in Conflict and Terrorism*, 30(6): 541–61.

Trocki, Carl (1979), *Prince of the Pirates: The Temenggongs and the Development of Johor and Singapore, 1784–1885*, Singapore University Press, Singapore.

Trocki, Carl (1990), *Opium and Empire*, Cornell University Press, Ithaca NY.

van Houtum, Henk and Ton van Naerssen (2002), 'Bordering, ordering and othering', *Tijdschrift voor Economische en Sociale Geografie*, 93(2): 125–36.

Vijayan, K.C. (2004), '3-nation patrols of strait launched', *Straits Times*, 21 July.

Wee, Vivienne (1985), 'Melayu: hierarchies of being in Riau', PhD thesis, Australian National University, Canberra.

Wee, Vivienne (2002), 'Ethno-nationalism in process: ethnicity, atavism and indigenism in Riau, Indonesia', *Pacific Review*, 15(4): 497–516.

Wee, Vivienne and Cynthia Chou (1997), 'Continuity and discontinuity in the multiple realities of Riau', *Bijdragen Tot de Taal Land en Volkenkunde*, 153(4): 527–41.

# INDEX

## INDONESIA UPDATE SERIES